D0115847

Fartology

The Extraordinary Science Behind the Humble Fart

Stefan Gates

Hardie Grant

QUADRILLE

Publishing Director: Sarah Lavelle
Designer: Luke Bird
Design Manager: Claire Rochford
Copy Editor: Kathy Steer
Editorial Assistant: Harriet Webster
Production: Vincent Smith and Nikolaus Ginelli

First published in 2018 by Quadrille, an imprint of Hardie Grant Publishing

Quadrille
52–54 Southwark Street
London SE1 1UN
quadrille.com

Reprinted in 2019 (twice)
10 9 8 7 6 5 4 3

ISBN: 978 1 84949 968 2
Printed in China

Fartology
Contents

Introduction   4

01: Fart Chemistry   10

02: Fart Biology   36

03: Fart Physics   82

04: Ask the Fart Doctor   104

05: Assorted Farty Trivia   128

Introduction

Every Fart Tells a Story

H ello. I know you're a little nervous but you're excited, too. You're about to go on an amazing journey of discovery, because this is the story of the natural beauty, extraordinary complexity and breathtaking ingenuity of your body. Of the tireless work of trillions of tiny creatures inside you, and of the triumph of evolution. It's a story of everything that food is: a delicious amalgam of chemistry, physics and biology born in the screaming hot plasma of the Sun, transmuted by photosynthesis, made physical by biochemistry, made flesh by metabolism and made human by pleasure and pain, sensory perception, love, guilt and embarrassment.

But step back from the magnificence of science for a moment because science doesn't care about you – it just *is*. It has no moral framework and no sense of duty – it just creates you and spits you out naked into a world built by its facts. Yet there is something about you that transcends those facts: your sense of self, your capacity for abstract reasoning, your ability to love, hate, believe and enjoy and, yes, to feel embarrassment when you let fly a cheeky guff. Life as a human pits us between the knowledge of science and the chaotic emotional self-awareness of being, and this is where farts have the power to thrill. To take uncaring science and shout loud and clear that you can understand it, take what it has to offer, spin it around your finger and every now and then flick it the bird.

Farts are the glorious, stinking klaxons of humanity, proclaiming loud and clear that you and I are vital, alive, flawed, complex, self-aware, created by science but triumphing over it, on one side trapped by a society that demands suppression of our true natures, yet on the other side set free by our desire to rebel. They are filthy-clean, cheeky-dirty, made of this earth, organic and complex, sweet and stinking, and they prove that we have real and simple beauty – oh, such beauty.

The poet Andrew Marvell said:

> *Let us roll all our strength and all*
> *Our sweetness up into one ball,*
> *And tear our pleasures with rough strife*
> *Through the iron gates of life.*

I'm pretty sure he meant: *understand your body and love your farts.*

A fart's a fart

Let's get something out of the way right now. A fart's a fart. If this were an academic medical book about flatus, it would be full of words such as eructation, emanation and fundament. But it's not. It's a popular science book about farts, smells and bums, designed to fascinate you, to make you fall in love with your amazing body and inspire you with science. Those formal words do crop up but if it makes things clearer, simpler or more interesting, I'm happy that a fart is a fart, a bum is a bum, and if necessary we talk about arseholes instead of anuses.

This is not meant to be a funny book about farts (there are plenty of those available if that's what floats your boat). This is a *fascination* book with three clear motives:

1. To make you fall in love with science.
2. To stop you feeling physical and social discomfort by keeping farts in when you'd genuinely benefit from letting them out.
3. To make you smile.

Why me?

I fart a lot. I still get a tiny bit embarrassed by my farts, but I'm working on it. I adore science, especially when it's blended with food – I'm a foodhead, a TV presenter and a food and science communicator. Together with my small team, I work my pants off creating huge stage shows that we perform around the world, taking complex science and making it fascinating, explosive and revolting. I've had gastroscopies, MRI scans, eaten PillCams and I've had liposuction to dig out the cooking ingredients from human fat. I love my farts and I hope you'll love yours, too.

A medical note

This is absolutely not a medical book and does not intend to give advice to those who suffer gastric problems such as irritable bowel syndrome (IBS), though they have my most heartfelt sympathy. See your doctor and don't take anything written here as a cure or solution.

Scientific sources

The scientific literature on farting is relatively thin on the ground and much is contradictory with widely differing methodology and study sizes. I've worked with gastroenterology researchers to make sure everything is accurate and we've trodden a path that takes into account differing results, but if you've found anything new that sheds light on farts I'm all ears.

Chapter 01:
Fart Chemistry

The basics: what is a fart?

Everybody farts. It's a completely natural, healthy part of your digestion, with a typical person farting around 10–15 times a day, producing an average of 1.5 litres/ 2½ pints of gas. We fart less at night but more as we start to eat a meal because of reflex actions from the stomach that kick-start the colon. Women fart less than men by volume but tend to produce smellier gas, and both volume and smell are closely related to the food you eat. Bigger, smellier farts are no more or less healthy than smaller, smell-free farts.

Around 25% of your fart gas is simply swallowed air that has gone all the way through your body, but the other 75% is made by various digestive processes, mainly the breakdown of dietary fibre by your gut bacteria. The best fart fuel comes in the form of complex carbohydrates, especially molecules called oligosaccharides (carbohydrates with 3–15 sugar units) mainly found in beans, root vegetables, onions, brassicas such as cabbages and cauliflowers, fruit and dairy products. The process is known as fermentation, metabolism, rotting or digestion, and it's anaerobic, meaning that it happens without oxygen (most of the 100 trillion microbes in your gut won't survive in oxygen-rich environments).

{ 'Just to be clear, you're carrying around 100 trillion little alien creatures in your colon that are not human.' }

Just to be clear, you're carrying around 100 trillion little alien creatures in your colon that are not human. There are huge differences in both gas volume and smell production among the population, with various studies showing ranges of 3–40 farts per day ranging from 400ml/14fl oz to 2.5 litres/4½ pints and a wide variety of gases and smelliness. That's due to natural differences in the types and amounts of bacteria that live in our colons. The colon (also known as the large intestine or bowel) is where the vast majority of gas-producing bacteria live.

Our farts are mostly nitrogen, carbon dioxide, hydrogen and occasionally methane. Some of the carbon dioxide is produced by acid-base reactions between the acidic gastric juices from your stomach and the alkaline secretions from your small intestine but the majority is produced by gut bacteria. Hydrogen is produced by bacterial fermentation but the nitrogen is mainly from swallowed air

SMELL COMBINATIONS

Over 99% of the typical fart is made up of substances that are totally odourless: nitrogen, carbon dioxide, hydrogen and occasionally methane. The smelly part is in that tiny 1% left over, and the smell may be made up of dozens or even hundreds of different compounds depending on your gut bacteria and your diet. You see, fart smells don't come from just one chemical compound – there'll be dozens, possibly thousands. Incidentally, strawberries have upwards of 30 compounds, but scientists working on cocoa aroma analysis have so far identified an incredible 20,000 molecules, and around 75% of them were previously unknown to science.

(the oxygen in the air is stripped out in the stomach and the first part of the small intestine). Some people, but not all, produce flammable methane, and that's due to special methane-producing bacteria.

The gases that make your farts smell are a tiny component, usually less than 1% of the volume, and the smells come from traces of hydrogen sulfide and several other compounds.

The digestive process is surprisingly slow. Although the first part of a meal could potentially get through to your colon (and theoretically start producing gas) within 2 hours, the digestion of an entire meal typically takes around 50 hours for adults and 33 hours for children, varying enormously depending on what you've eaten and how your body functions. Food usually takes around 4 hours to pass through your stomach, 6–8 hours to pass through the small intestine (longer if there's a lot of fat in it), and then everything slows down enormously through the large intestine, which takes about 40 hours. There's a big difference between men and women: men's digestion in the colon alone averages 33 hours and women's averages 47 hours.

Is a fart just a gas version of a poo?

We all know the real underlying question here: if I smell someone's fart, am I basically inhaling their poo? And if so, should I nip off somewhere to be sick? The answer is 'no'.

Well, maybe just a little bit. But first things first: what is a poo?

Poo

Poo (or *faecal matter* if we're being formal) is fascinating stuff – the phrase 'metabolic waste matter' barely does it justice. Everyone's poo is different but a typical day's poo weighs in at 100–225g/4–9oz and is usually 75% water and 25% solid matter. There's dietary fibre, lots of bacteria (both dead and alive) and lots of other compounds (see page 72 for a full breakdown).

Fart

In contrast, a fart is almost entirely gas: nitrogen, hydrogen, carbon dioxide, occasionally methane together with traces of the same flavour volatiles: hydrogen sulfide, methanethiol, indole, skatole and dimethyl sulfide. So, on a broad analysis you'd think that poos and farts couldn't be more different.

Let's return to the question 'Is a fart just a gas version of a poo?' Bacteria are so small (0.2 to 10 microns – viruses are even smaller) that some of them could *potentially* shoot out and become airborne alongside other gases when they are farted out of your anus. Some bacteria can survive as they float in the air, for instance, the mycobacterium *Tuberculosis* is an airborne bacteria that can float alive for hours (although tuberculosis is usually spread through the air when an affected person sneezes, coughs or talks – not by farts). Most airborne bacteria are likely to quickly die from dehydration or UV radiation, but yes, you *could* inhale the bacteria from someone's farts.

We do, of course, inhale the airborne gas components of other people's farts, but there's unlikely to be much solid matter with it. *However* (and it's a big however), the gases do interact with the smell chemoreceptors in the olfactory bulb behind your forehead so, in a funny way, a few molecules of someone else's fart do *become part of you*, but only for a while.

So where's the evidence?

There's precious little research in this area but the *British Medical Journal* did republish an item from *The Canberra Times* about a Dr Karl Kruszelnicki working in Australia who was asked by a nurse whether it was alright to fart in the operating theatre. He didn't know, so together with a microbiologist he asked a colleague to fart on to a petri dish from a distance of 5cm/2in – once fully clothed, and once with his trousers down, to see what would happen. The next day they checked the petri dishes and found that the one farted on through clothing had grown no bacteria, but the dish farted on naked had developed two visible lumps of (harmless) bacteria – ones usually found only in the gut and on the skin.

So farts are very different from poo but yes, farts can *theoretically* contain tiny traces of bacteria. Clothing does seem to act as a filter to contain them. What can we learn from this? If you're 5cm/2in away from the source of an unclothed fart, let's face it, you're probably too close.

Why do farts smell?

We all love the smell of our own farts (come on, you do really), especially when we're in a confined space or enjoying a Dutch oven* under the duvet. But where does that stench come from?

The journey of your farty smells starts when hard-working bacteria create highly volatile gases as they digest any remaining food in our colon. It's an amazing process called *metabolism* during which complex molecules are broken down into simpler ones (*catabolism*), and new molecules are created (*anabolism*) – and lots of gas is made.

* A Dutch oven is a fart performed in bed with the duvet held tightly over your head. Performing one on a friend or partner is funny but dangerous, and has been cited as one reason why Ethel Merman and Ernest Borgnine's celebrity marriage failed.

Most fart gas is completely odourless (neither nitrogen, carbon dioxide, hydrogen nor methane have a smell), but much more interesting is the tiny trace of flavour volatile gases that farts contain. What's extraordinary about them is that they aren't just terrifically smelly – they're *volatile*, which means they evaporate into a vapour very easily**, and that's how they manage to float around in the air and make their way up our noses.

Every fart will contain different compounds produced as by-products of breaking down food, often from protein found in high concentrations in meat, nuts, seeds and beans. The most pungent farts are often made by the breakdown of amino acids (the building blocks of protein) in our diet, which are found in lots of food but especially in beans, cheese and meat. These aren't necessarily the fartiest foods by volume, but usually the fartiest by smell.

** This is one of those ideas that scientists often leave unexplained to the layman, causing an infuriating fact-blockage to everything that follows it. I'll explain. The states of matter follow this order: 1. Solid 2. Liquid 3. Gas. 4. Plasma. But it isn't as clear as it seems because water can exist as a gas even though it's well below boiling point. For instance, when you put your wet towel in the sunshine, the water molecules will slowly evaporate until it's dry. Water molecules escape from the towel by bouncing around each other due to the energy they contain, until by chance one whacks into the others so hard that it gains enough energy to escape from the towel, becoming a gas even though it's lower than the 100°C/32°F boiling point of water. Like Brownian motion it's a random action but because it happens on such a vast scale over trillions of molecules, it becomes predictable. Substances that evaporate into vapour readily are *volatile*: they escape from their parent substance easily to zip around the air.

The WORST fart smells:

1. hydrogen sulfide =   rotten eggs
2. methanethiol =   putrid cabbage
3. trimethyl amine =   fishy
4. methyl thiobutyrate =   cheesy
5. skatole =   cat poo-ish
6. indole =   flowery dog poo
7. dimethyl sulfide =   cabbage
8. thiols =   eggy

The journey from bum to nose is pretty simple: after you've farted, these volatile gases get airborne and make their way into the air around us by virtue of Brownian motion (basically there are billions upon billions of molecules invisible to the naked eye in the air all around us, and they are constantly moving around and bumping into each other and spreading from their source in a random movement). It happens on such a huge scale that gases spread out relatively evenly into the atmosphere from bum to nose and beyond.

Next comes *olfaction*. As you breathe, those flavour volatiles get sucked in and head up the nasal passages until they find mucus in the olfactory epithelium, which is connected to your olfactory bulb (which sits by your brain behind your forehead). Some of the molecules are dissolved into mucus, which flows constantly and is replaced every 10 minutes or so. The flavour molecules that are dissolved into the mucus are then detected by olfactory chemoreceptors (little structures that detect smelly chemicals), which send a tiny electrical signal via a postage stamp-sized patch of neurons to your olfactory bulb and then via axons (imagine extraordinarily tiny little electric cables) to your brain, which then interprets the signal and provides you with the sensation of fart smells.

There's a little mystery here: no one knows for sure *how* flavour molecules actually interact with smell receptors to fire off a particular message identifying a unique smell. It used to be thought that there was a 'lock and key' concept at work: each flavour molecule was like a key and it fitted into a specific chemoreceptor lock. The trouble is that there are around a trillion different flavour compounds we can smell so we'd need a trillion different chemoreceptors, which seems unlikely. We know that there are also 400 genes that code for smell receptors.

PRACTICAL ADVICE

I'll often find myself needing to let fly a gust of wind in the middle of the night but unsure whether its flavour will be fragrantly sweet or thunderously sour and dyspeptic. Out of respect for my wife (who values her sleep highly), I will usually cup my hand and let rip into it so that I can direct it away from her delicate nose (she does have an extraordinarily fine sense of smell) and instead towards my own. Thinking about it, perhaps I'm just rubbish at sharing.

The smells

Fart Gas	Smell	Notes
(or flavour volatile)	*(with my personal marks out of 10 for rottenness)*	
Hydrogen sulfide (H_2S)	Rotten eggs *9/10*	At high concentrations it's very poisonous, flammable, corrosive and explosive. The product of many types of dietary fibre, egg whites also contain lots of ovalbumin, which breaks down and reacts with hydrogen to form hydrogen sulfide. Only people with sulfate-reducing bacteria in their gut can produce this gas, which is thought to be about 50% of the population.
Dimethyl sulfide (CH_3SCH_3)	Variously described as a mixture of baked beans, cabbage, sweetcorn and rotting meat *7/10*	Sometimes used in petrol refining, food flavouring and paper production.

Methanethiol (also known as methyl mercaptan) **(CH_3SH)** | Skunky, rotten eggs, slightly garlicky *7/10* | Like dimethyl sulfide it can be added to household gas to make odourless methane 'smellable'. So strong that you can smell it in concentrations as low as 10 parts per billion.

Trimethylamine (C_3H_9N) | Rotten fish, petrol, ammonia, household gas *8/10* | Used in the food flavouring industry for added fishiness and even porkiness.

Methyl thiobutyrate ($C_5H_{10}OS$) | Cheese, egg, sulfurous *7/10* | Found in strawberry aromas. Sometimes added to food to enhance cheesy, tomatoe-y, fruity and savoury flavours.

Skatole (also known as 3-methylindole) **(C_9H_9N)** | Petrol, gas, cat poo *7/10* | Thought to be one of the US military's secret non-lethal weapons. We make it from the amino acid tryptophan as we digest foods, especially meat, egg white, soya beans.

Indole (C_8H_7N) | New rubber football, slightly floral, dog poo, general animalistic flavour, orange blossom *7/10* | As with skatole, it's produced from the breakdown of the amino acid tryptophan in our guts. It's used in the perfume industry, especially in the production of synthetic jasmine oil.

What's the difference between men's farts and women's farts?

A glorious piece of research published in the journal *Gut* (1998: 43:100–104) is clear that women's farts are much smellier than men's (if anyone spotted my air-punch just now, I can only apologise). Not just a mere sniff smellier, but a whole LOT smellier due to higher concentrations of the prime fart-stench gases. Women's farts were found to have a 200% higher concentration and 90% higher volume of hydrogen sulfide than men's, as well as 20% higher concentration of methanethiol. And when offered to two experienced fart judges (yes, it's a job), the women's farts were judged to be clearly more offensive than men's. Admittedly the study was a small one involving 16 volunteers, but this tiny field of research is renowned for small sample sizes and few studies. Come on – 200%! Go, girls!

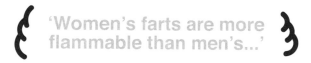

{ 'Women's farts are more flammable than men's...' }

The same cannot be said for the *volume* of women's farts. This particular study showed that men clearly trumped the women both in the average size of individual farts (118ml/4fl oz) compared with women's (89ml/3fl oz), but also the sheer number of farts:

men produced more than women by a ratio of 52:35. Go boys!

Admittedly, Tomlin, Lowis and Read's 'Investigation of normal flatus production in healthy volunteers' found that 'Women and men expelled equivalent amounts.' Boo. Clearly this is an area that needs more a lot more research. Come on folks, this is your chance of a smelly Nobel prize.

Women's farts are more flammable than men's as they produce more methane due to their higher proportions of methane-producing bacteria. About 60% of women produce significant amounts of methane but only 40% of men. This may also be why women produce less gas overall, as methane-producing bacteria use so much hydrogen to fuel methane production. Hormone replacement therapy (HRT) is reported to cause greater gas production due to progestogen, which slows the transit of food through the intestines, allowing more of it to be converted to farts.

Can you catch your farts in a jar?

Really, what's the point of collecting your farts? Well imagine how wonderful it would be if we had a National Fart Bank? A vast high-tech catacomb built to survive nuclear attack and stacked with cryogenically preserved farts from throughout history. A bit like the Millennium Seed Bank but less… seedy. How much richer would our world be if the rectal gases of the great and the good (and maybe the little and the bad, too) were preserved for posterity? Imagine if, rather than pondering the Dissolution of the Monasteries, kids could sniff the guffs of Henry VIII, Cardinal Wolsey or Catherine of Aragon? I bet Julius Caesar's farts were cat-poo flavoured and that Cleopatra's smelt a bit like damp spring pavements. Da Vinci's definitely niffed of oregano. Whose would you like to smell? Pocahontas's? Genghis Khan's? Jesus's?

There is one fascinating problem in the long-term storage of smells, but first let's find out how you'd go about starting a fart library of your own. It's dead simple and it's all down to Archimedes' principles about the displacement of water. All you need is to be sitting on a fart that's ready to blow, and to have access to a bath and a jam jar with a decent lid. First, run yourself a lovely bath without bubble bath, then pop your clothes off, grab your jam jar and hop in. Lie down on your back then take the lid off the jam jar and immerse it in the bath until it's full of water. Turn the jam jar upside down, still full of water and still under water, and hold it above your bumhole.

When you're in position, let rip with as much gas as you can eject (we're not at home to Mr Shart, mind, this is a family show), and it should bubble up into the jar, displacing a weight of water equal to the magnitude of its upward buoyancy. Basically, the air pushes up, forcing water down and remains trapped in the jar. Now place the lid on the jar tightly, and turn it the right way up. The water will fall to the bottom and the fart rises to the top. Bob's your fart. In the eighteenth century this was known as 'collecting a gas by downward displacement of water'. If you want to analyse the fart, just submerge it in a large body of water, turn it upside down before opening, then extract the gas using a syringe.

How long do your farts stay smelly?

So, we've got the practicalities of fart collection down, but there is a cloud looming over our National Fart Bank. Longevity. How long do those farts last? I discussed this at length over a kick-ass vegetarian thali with Professor Andrea Sella, a chemist notorious throughout the great University College London chemistry department for his explosive lectures and fast-and-loose attitude to health and safety. I love him dearly and when you meet him, you will too.

The problem is that after trapping your fart in a jar, several problems crop up that can interfere with the captured smells. You need to remember that those smells are made up of volatile molecules with very specific odour profiles and may be highly reactive. If they react with other molecules around them, the interaction may create completely different molecules with a different smell or no smell at all – for instance hydrogen sulfide could eventually react with other components of your fart or the air to form a third product.

The threats to our fart flavours are:

1. Reactions with the water vapour in the fart (or captured in the jam jar and bath water), causing smell compounds to dissolve in the water.
2. Oxidation (the gaining of oxygen, or loss of electrons in a reaction), especially caused by exposure to ultraviolet light, which interacts with organic molecules and can cause oxidation of the ephemeral nature of the fart.
3. Reactions with other gases in flatus.
4. Reactions with the container materials. Glass is pretty much inert so not too much of an issue, but metal lids and plastic seals may be problematic.

We should seize the opportunity to smell farts just as we 'seize the day'. *Carpe flatus*! So my National Fart Bank idea is not

An old married couple are in church one Sunday when the woman turns to her husband and says, 'I've just let out a really long, silent fart. What should I do?'

The husband turns to her and said, 'Replace the battery in your hearing aid.'

without problems but if the gases are cooled enough perhaps in liquid nitrogen at -196°C/-320.8°F or even liquid hydrogen at -253°C/-423.4°F the reactions will slow down so much and the smells are preserved. And Bingo! The National Fart Lab's a goer. Incidentally, must dash – that thali's kicking in...

Do animals fart?

Dani Rabaioitti and Nick Caruso are fantastic ecologists. Together they have tackled one of the great questions of our time:

'Does it fart?', harnessing the power of Twitter to galvanise animal researchers across the world to pool their knowledge. They have very kindly agreed to let me share some of the findings they published in their very excellent book *Does It Fart?* (Quercus Books, 2017) and in return, I promised that you'd all buy a copy. Hope that's okay?

Animal	Does it fart?	Notes
Herring	Yes	They gulp air and then fart it out for communication purposes.
Goat	Yes	In 2015 a plane carrying 2000 goats was forced to land due to the copious gas they produced.
Cow	Yes	They do fart but create more gas as burps. Cattle produce a third of all agricultural greenhouse gas emissions with around 250 litres/55 gallons) of methane per day from total gas emissions of 600 litres/132 gallons) a day. That's HUGE.
Kangaroo	Yes	Less than cows, but similar to horses.
Portuguese man o' war	No	They don't have an anus. They eat by liquefying food with digestive enzymes.

Spider	Nobody knows	Most spider digestion happens outside their body using venom with enzymes.
Elephant	Yes	They have huge digestive tracts to break down tough foods like tree bark.
Bird	No	Birds don't have the gut bacteria for the job, and have a very fast digestive process.
Termites	Yes	Termites are tiny but there are so many on earth that they produce 5–19% of all methane emissions.
Goldfish	No	Although they do have gas-producing bacteria in their gut, they are more likely to burp.
Woodlouse	Kinda	Woodlice convert nitrogenous waste into ammonia and excrete it in bursts, usually lasting a few minutes, but sometimes an hour or more.

How to build a stink bomb

Stink bomb ingredients used to be in every kids' chemistry set, but these days we are much more cautious about letting children get their hands on some of the more dangerous components, and with good reason.

There are lots of ways to make a stink bomb, from dead simple to highly dangerous. This method is middling-dangerous because it uses matches and ammonia, which can be toxic and corrosive. If you're under 16 you must get help from an adult, and if you're an adult, you must supervise anyone under 16 very closely. The stink bomb will need to be stored somewhere for several days as it brews, and it's vital that it isn't found by kids, pets or any unsuspecting adult.

Warning:
1. Ammonia can be toxic and corrosive.
2. The ammonium sulfide is flammable and toxic in high concentrations.
3. Label the bottle VERY clearly so no one drinks or uses it for the wrong purpose.
4. Don't drink, throw or spill the resulting muck. It's revolting.

What's going on?

You're going to mix the sulfur from match heads with ammonia (a cleaning product), and they react to make ammonium sulfide, which smells very much like rotten eggs. The chemical reaction is:

$$H_2S + 2\,NH_3 \longrightarrow (NH_4)_2S$$

You will need:

- Box of matches
- Ammonia sold for household cleaning
- Empty plastic 500ml/17fl oz bottle
- Wire cutters OR pliers OR strong scissors

Method

1. Cut all the heads of the matches using the wire cutters or scissors, and put them into the empty bottle.
2. Add 30ml/1fl oz ammonia.
3. Squeeze the bottle a little then put the top on tightly (so that the gas doesn't over-pressurise the bottle).
4. Seal the bottle and shake it gently to mix.
5. Leave in a safe, secure place for 3–5 days for the reaction to happen.
6. Uncap the bottle and use the ingredients OUTSIDE to stink your friends out but DON'T get it on clothes or carpets.

Chapter 02:
Fart Biology

What's the journey from food to fart?

I t all begins in that vast sphere of hot chaotic plasma we call the Sun, a yellow dwarf star weighing 330,000 times the mass of Earth, made up mainly of hydrogen (73%) and helium (25%), and with around 4.6 billion years on the clock. The Sun generates energy by the nuclear fusion of hydrogen nuclei into helium, and one of the by-products of this is light. Photons of light radiate from the Sun in electromagnetic waves, with only a tiny proportion of them reaching Earth – but it's the perfect amount to sustain our ecosystem.

Nuclear fusion

When the photons of light reach Earth some of them will land on plants containing the pigment chlorophyll, kick-starting the miraculous* process of photosynthesis which provides virtually all the energy used by living things. Photosynthesis really is amazing: it uses light energy, water and carbon dioxide to create chemical energy that's stored as carbohydrate molecules in the plant, some of which we can eat as food, and at the same time produces oxygen as a waste product. That oxygen is the oxygen you and I are breathing right now.

Normally photosynthesis is invisible, and you just have to take my word that it's going on, but if you want to see it in action there's a beautiful demo you can do to make it all plainly visible. Buy some pond weed from a pet fish shop – the best one I've found is Canadian

---

* It's actually the opposite of a miracle – it's solidly understood science, but I just get *really* excited by it.

pondweed (*Elodea canadensis*). Cut the top 1cm/½in of the weed off, using sharp scissors, put it in some water and weigh it down so it's underwater (a bulldog clip works well). Now shine a light at it. Soon you should be able to see little bubbles popping out of the top. That's oxygen being produced. I know, I *know*! I've made a video of it, so do please do take a look at the YouTube channel GastronautTV.

Go deeper into photosynthesis

$$6CO_2 + 6H_2O = C_6H_{12}O_6 + 6O_2$$

Go on, take a moment to go deeper. The reaction is simple on paper: plants capture light and use it to stitch together 6 molecules of carbon dioxide (CO_2) and 6 molecules of water (H_2O) into a molecule of glucose sugar, $C_6H_{12}O_6$. But in reality it involves many steps. First a coloured light-capture molecule – chlorophyll – absorbs the light and out pops an electron so energetic that it can split water – a bit like electrolysis. This sets off a cascade of chemistry where the electron is handed from one molecule to another to milk away the energy and store it into molecules called adenosine triphosphate (ATP) that every organism uses to drive metabolism. That same ATP is then used in a second cascade of steps to convert CO_2 into glucose, the sugar of life.

There's a lovely statistic about the total energy rate captured across the planet by photosynthesis at any one time: 130 terawatts. It's hard to get your head around but that's only 0.1% of the solar energy that strikes the Earth (it's also three times humanity's total energy consumption).

What's food and what's not?

You may be surprised to know that *all* matter that fits through your digestive tract is, strictly speaking, edible, regardless of whether it's bread, wood, grass, chewing gum or chunks of iron crowbar. If you can't get any nutrition from it, as is the case with the crowbar and the chewing gum, it's known as insoluble fibre (useful for helping to push everything through your intestines), and if you can't digest it in your small intestine but can break it down in your bowel (as is the case with the wood and grass) it's called soluble fibre. Everything else is considered a nutrient, and you usually benefit from digesting it in some way.

How does digestion work?

You've identified some nice edible food, possibly even cooked it yourself. Now it's time to digest it. The average amount of time for your body to digest food is a whopping 50 hours in adults (33 hours in children), with the large intestine alone taking 40 hours. Now the fun begins…

Digestion Stage 1: Chewing – mechanical breakdown

Your molars exert a vast force of up to 120kg/265lb as you bite down on your food, and you'd be forgiven for thinking that this mechanical breakdown with its crunching, crushing and squashing, is the most important part of the digestive process, but you'd be wrong. All you're really doing is increasing the surface area of the food so that the next stage of digestion can happen. That said, the most dramatic and enjoyable part of eating is the sensory experience in your mouth, *not* the act of getting nutrition out of it. As we chew our food we enjoy the smell and taste of the flavour molecules as they interact with our chemoreceptors, and we enjoy textures as the food activates our mechanoreceptors. We'll also sense the heat via our thermoreceptors and even the sound of our food (especially crunchiness) as it interacts with our aural receptors (sound perception is actually a hypersensitivity to touch). But none of these features of food has nutritional value other than to encourage us to enjoy foods with lots of energy. Interestingly, the process of eating a meal prompts your nervous system to send messages to your large intestine that encourage you to go to the toilet, which is odd as the food you're eating is still hours away from getting anywhere near your colon. At least it knows that more is coming.

Digestion Stage 2: Saliva – enzymatic breakdown

You produce about 2 litres/3½ pints of saliva a day, containing 94–99.5% water alongside an enzyme called *amylase*, which starts breaking down your food into its component parts. (Saliva also contains small amounts of calcium, fluorine, magnesium, sodium, uric acid, proteins, peroxidase and bacteria.)

Try this easy demo:

Mix up some instant custard and split it into two glasses. Spit into one glass 4–5 times (I usually get several people to spit into it, which they find surprisingly revolting) and stir it in with a teaspoon, then pour both custards down on to a chopping board held at an angle. The spit-mixed custard will be very watery while the control custard will still be thickly viscous. The saliva has broken down complex sugars in the custard, making it very watery, and the reaction happens very quickly.

Digestion Stage 3: Swallow food via the oesophagus

Around 50 pairs of muscles are involved in the complex process of preparing and swallowing your food. As the food is pushed towards the back of your mouth a swallowing response is triggered that pushes it towards your throat and at the same time triggers a reaction that closes your voice box (*larynx*) and stops your breathing so that the food doesn't go down your windpipe (*trachea*). Then a series of muscle contractions called *peristalsis* pushes the food down towards the bottom of your food channel (*oesophagus*) until it hits your oesophageal sphincter, which looks like a cat's bum (mainly because it sort of... is). It's a valve to let food into the stomach but keep the gastric juices out – unless you vomit. It also opens to allow burps out.

Digestion Stage 4: Stomach – acid and enzymes

This muscly organ sits to your left side and is a tiny 75mm/3in capacity bag the size of a fist when empty, but expands to 1 litre/ 1¾ pints capacity normally and up to 2 litres/3½ pints or more should it need to. As soon as food enters the stomach, its lining secretes gastric juices, which include digestive enzymes called *proteases* that break down protein. It also produces gastric acid, including a fair amount of hydrochloric acid which kills bacteria, denatures proteins and effectively recooks your food. When you vomit, that acrid taste in your mouth is the delicious tang of these gastric juices. The stomach churns everything around to mix these enzymes and acids together with your food. Food stays here in the stomach from anything from 15 minutes to 4 hours. The more fat you eat, the longer food will stay in your stomach, allowing more time for you to create fat-digesting bile. The stomach doesn't absorb many nutrients itself (that's mostly left to the next stage of digestion) except for some medications, amino acids, some alcohol and caffeine.

Weirdly, you have some taste receptors in your stomach that can send a sensation of pleasure to your brain when they interact with glutamates, sugars, carbohydrates, proteins and fats. The stomach slowly uses peristalsis to push the mixed-up food and gastric juices (now called *chyme* by gastroenterologists) into the next stage of the journey via the *pylorus* – another sphincter-like opening – and into the first part of the small intestine called the *duodenum*.

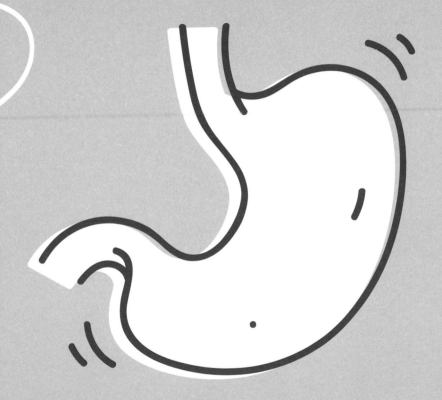

Digestion Stage 5: Small intestine

The small intestine is very long but very thin. It's where most of the nutrients from our food get absorbed, and lots of substances need to be added to the food to make that happen. Bile is produced by the liver and added via the gall bladder to break down fats. Pancreatic juice containing lots of enzymes is added via the pancreatic duct. Interestingly it's highly alkaline as it contains lots of bicarbonate ions, and this neutralises the acid from the stomach's gastric juices, creating the right acidity levels for the enzymes to work properly, but also creating some fizz: carbon dioxide gas from the resulting acid-base reaction.

Food usually spends 6–8 hours passing through the small intestine, squeezed through by continuous worm-like wave peristalsis contractions. The length of the small intestine alone averages 3–5m/10–16ft but can be as short as 2.75m/9ft and as long as 10.4m/34ft. The surface of it looks like velvet, a short fur made up of microscopic little fingers called *villi* and *microvilli* that offer a huge surface area of around 30m²/323ft² for nutrients to be absorbed through.

The molecules in your food are broken down into smaller, more usable parts, such as vitamins and minerals, sugars (from carbohydrates), amino acids and peptides (from protein) fatty acids and glycerol (from fats). These pass through the walls of the intestine and into the adjacent blood vessels. Anything that's left passes through another valve called the *ileocecal valve* that stops poo in the colon from going back into the small intestine.

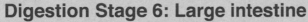

Digestion Stage 6: Large intestine

The large intestine is also called the *colon*, bowel or large bowel, just to confuse you. It's 1.5m/5ft long – nowhere near as long as the small intestine – but is much wider, hence its name. This is where the whole digestive process slows down, giving the bacteria time to work their magic. Anything that hasn't been absorbed into the blood in the small intestine ends up here, ready to be broken down, used or ejected.

The large intestine has the shape of an incomplete square, starting with the *ascending colon* travelling upwards on your right side followed by a 90-degree turn to the *transverse colon* going from right to left above your belly button, then another 90-degree turn into the descending colon on your left side. There's a short turn again back to

the middle before heading down to the holding chamber called your *rectum*, and then finally below that to your *anus*.

The large intestine is where all the flatulent magic takes place, as well as several other processes including compaction and dehydration. There are about 100 trillion microbes in your gut weighing about 200g/7oz, made up of over 700 species of bacteria as well as other microbes such as fungi and protozoa. They live and reproduce in our colon and mix together with chyme and mucus to turn the chyme into faeces. We don't absorb many nutrients here, but we do get water, vitamins created by our bacteria, thiamine and riboflavin. The bacteria also break down fibre to create fuel for themselves and produce short-chain fatty acids. The big news is the bacterial breakdown of our soluble fibre (mainly indigestible carbohydrates), which results in fart production.

Digestion Stage 7: Rectum

The last section of the large intestine is about 12cm/5in long and it's a storage chamber for gas and faeces. As it receives faeces from the large intestine it expands and the pressure on the stretch receptors stimulate you to feel the need to go to the toilet. When the rectum is full, the pressure will force the anal walls apart and the faeces will enter the anal canal, shortening the rectum and pushing the faeces out with the body's last set of peristaltic waves.

Digestion Stage 8: Anus

Nearly there. The anal canal is around 2.5–4cm/1–1½in long and points down and a little backwards, leading to the anal opening, which is controlled by two circular muscles called the internal sphincter, (which you have no control over), and the external sphincter (which you do).

Stage 9: Now wash your hands!

What are the fartiest foods on earth?

These top tips for increasing your fart harvest were compiled with the advice of gastroenterology experts and scientific studies, but also with the help of hundreds of lovely people who responded to my surveys and requests.

Jerusalem artichokes

Probably the most powerful fart-producing fuel in the world due to high proportions of a carbohydrate called inulin (up to 75%), and the peculiar characteristics of its activity in your bowel. Jerusalem artichokes are *so* farty that I've given them their own mini-chapter (see page 54).

Beans and other raffinose-rich foods

St Jerome (AD 347–420) advised nuns to avoid beans because in *partibus genitalibus titillations producunt* ('they tickle the genitals'). Quite how St Jerome had such intimate knowledge of nuns' genitals is anyone's guess, but I assume he was mistakenly referring to flatus. Beans such as soya beans, pinto beans and kidney beans, as well as broccoli and asparagus, are rich in fibre, especially the clever oligosaccharide *raffinose* as well as stachyose and verbacose. We don't have the enzyme needed for breaking them down in our small intestine (an enzyme called a-GAL) so it passes pretty much intact into our colon. Happily, the bacteria in our colon do possess the enzyme, and they ferment it with a special gusto, producing a huge amount of gas partly because they love fibre, but also because (as with inulin) they are fuelled by it to work harder at the same time, effectively turbocharging the process (the oligosaccharides work as a prebiotic fuel).

Onions, garlic and leeks

These contain fructans, which are polymers of fructose sugar, another complex carb that needs to be broken down by gut bacteria rather than by enzymes in the small intestine. Different onions produce different volumes of fart.

Cruciferous vegetables

Also known as brassicas, these include cabbage, cauliflower, broccoli and Brussels sprouts. They contain lots of soluble fibre for bacteria to feast on (and lots of vitamin C), but they also contain lots of glucosinates – organic compounds derived from glucose and amino acids that contain sulfur and nitrogen – which give brassicas their slightly bitter flavour and which can break down into smelly sulfurous compounds.

Wholegrains

Including wholegrain bread, bran, oats, etc. Packed with soluble fibre that's delivered straight into your colon, you can almost see your gut bacteria rubbing their hands with glee. These foods may create gas volume, but are unlikely to contribute much smell to your flatus.

Fruit

You may not have expected to see these here, but many fruits (and especially dried fruit) including pears, apricots, prunes and peaches contain natural sugar alcohols among which bacteria ferment very happily in our bowels.

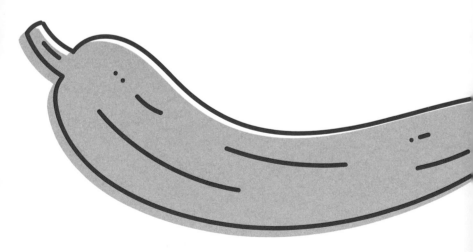

Unripe bananas

Unripe bananas have more starch (complex sugars) and less simple sugars than ripe bananas, and these *resistant starches* pass straight to the colon for digestion. They aren't dangerous, although why anyone would want to eat an unripe banana is beyond me.

Orange pith

This is rich in pectin, another starch (complex sugar). As with unripe bananas, this heads directly south for fermentation in the colon.

Meat and dairy products

High-protein food doesn't necessarily increase the volume of your farts (it usually produces less), but can add significantly to the smell. Protein is based on amino acids, two of which can break down into sulfur-based stinky compounds*. One study shows that the amino acid cysteine can increase hydrogen sulfide emissions in the gut by up to 700%.

Fatty foods

Fat is fascinating. It is broken down into low pH fatty acids in the small intestine, so we release high pH alkaline bicarbonate to neutralise it, bringing it back to a neutral pH nearer 7 (clever, aren't we?). The acid-base reaction between the bicarbonate and fatty acids releases carbon dioxide gas, which can create an uncomfortable bloating feeling (try adding a teaspoon of bicarbonate of soda to 2 teaspoons of lemon juice and you'll get the picture). Some of this carbon dioxide is absorbed into the blood, but some will make its way to the colon.

---

* The two amino acids containing sulfur are methionine $NH_2CH(CH_2CH_2SCH_3)CO_2H$ and cysteine $NH_2CH(CH_2SH)CO_2H$. The latter gets oxidised to make $CH_2-S-S-CH_2$ 'disulfide bridges' that keep proteins in the correct shape.

Potatoes and cereals

These contain fart-producing fructans that are broken down in your colon, but they also have an interesting twist to them: they can actually *reduce* smelliness – one study showed that eating them caused hydrogen sulfide production to be reduced by 75%.

Chilled pasta and potatoes

This is a weird one. Cooked starchy foods develop large carbohydrates called resistant starch 3 as they cool (especially in the fridge), and this is another complex carb that gut bacteria love.

Lactose

This is the sugar that gives milk its slight sweetness, and it can increase your fart production, especially if you're lactose intolerant. If so, you lack the enzyme needed to break it down in your small intestine, so instead, it's broken down in your colon by bacteria. The gut bacteria that dine on the undigested lactose produce hydrogen gas – that's one of the diagnostic signs of lactose intolerance.

Shiitake mushrooms

These contain a polyol (sugar alcohol) called mannitol that's digested by gut bacteria and acts as a mild laxative. It's used as a diabetic food sweetener specifically because it is poorly absorbed in the small intestine, and that leaves it to be broken down in the gut.

Body-building protein powder

Body builders are notorious for having smelly farts, and this is thought to be due to the extra sulfur-producing protein they consume, which causes hydrogen sulfide emissions from gut bacteria to increase.

Sorbitol-sweetened sugar-free gum

The sugar substitute sorbitol is usually made from corn syrup and is lower in calories than ordinary sucrose (table sugar). It's often used to make your chewing gum taste sweet. It cannot be digested but it can cause gastrointestinal distress and can be used as a laxative, drawing water into the large intestine and stimulating bowel movements. Its large molecular weight is thought to stop it being broken down in your small intestine so it ends up in your colon, where it can increase the fermenting ability of your gut bacteria, causing gas production.

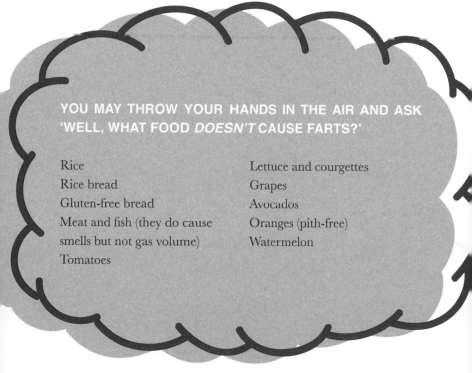

YOU MAY THROW YOUR HANDS IN THE AIR AND ASK 'WELL, WHAT FOOD *DOESN'T* CAUSE FARTS?'

Rice
Rice bread
Gluten-free bread
Meat and fish (they do cause smells but not gas volume)
Tomatoes

Lettuce and courgettes
Grapes
Avocados
Oranges (pith-free)
Watermelon

What on earth are artichokes?

The early English adopter John Goodyer swore that 'they stir and cause a filthy loathsome stinking wind, thereby causing the belly to be pained and tormented'.

What are Jerusalem artichokes?

Jerusalem artichokes are directly responsible for creating high-volume flatulence (this depends on your own unique biome profile but I for one pump like a steam train after eating them). Without question the fartiest foods on earth, these knobbly tubers do not originate from Jerusalem and are not related to artichokes (although the taste is similar). They are a type of sunflower native to North America where they're marketed as 'sunchokes' (though my favourite name is the French *topinambour* from the allegedly cannibalistic Amazonian *Tupinambas* tribe). They are easily mistaken for ginger or turmeric, and are a beast to get rid of if you've made the mistake of planting them in your garden. They grow and reproduce rapaciously, and after realising the sheer impossibility of matching supply with consumption (my wife having refused to eat them any more), it took me four years of frantic digging to clear them from my allotment.

Why do they make you fart?

The blame lies firmly at the door of *inulin*. It's a large and relatively complex carbohydrate molecule (specifically a polysaccharide*, made up of long chains of monosaccharide units) which you can clearly taste on your tongue (crunch a slice of raw Jerusalem artichoke and you'll experience a delicate apple-like sweetness) but which we lack the right enzymes to break down in our small intestine. In fact it's so difficult for our gut enzymes to break it down that Jerusalem artichokes are considered 'indigestible'. So the inulin travels pretty much intact to the colon where the resident bacteria feast upon them and produce vast quantities of gaseous by-product including carbon dioxide, hydrogen and methane. The fascinating twist is that it's also converted into a prebiotic nutrient, and this supports all the other bacteria in your colon, helping them to thrive and work at full capacity, thereby boosting your gas production across the board. It truly is a wonderful thing. Inulin is in the artichoke as an energy store instead of starch, which is more commonly found in other tubers such as potatoes. Incidentally, inulin is also found in lower concentrations in wheat, bananas, onions, asparagus and chicory (the latter being the main industrial source of it). It's cropping up more and more these days as a low-sucrose food additive that can help calcium absorption (but you might want to avoid it if irritable bowel syndrome (IBS) causes you problems).

* Saccharides are types of sugar molecule and 'poly' in polysaccharide means 'many', which is why it's made up of monosaccharide units.

The history of the Jerusalem artichoke

There was a moment in the mid-1600s when it could have become
the new resilient root crop due to ease of growth, high yield and
the fact that its tubers can stay in the ground indefinitely. But in the
event it was ousted by that other newfangled product of the great
Columbian Exchange, the potato. Jerusalem artichokes have often
been grown as animal feed, and the French have had an on-off
relationship with them, switching them from human food to animal
food and vice versa for many years. Modern recipe writers invariably
fail to warn against these side-effects when they suggest you add
them to a dish but it was clearly noted by John Goodyer, quoted in
Gerard's Herbal in 1621:

> *'Which way soever they be dressed and eaten, they stir and cause a filthy
> loathsome stinking wind within the body, thereby causing the belly to be
> pained and tormented, and are a meat more fit for swine than men.'*

Can you stop them making you fart?

No. The thing is, there are good reasons to eat artichokes – they're delicious roasted, excellent raw in salads and they make a fine soup because that cheeky inulin is wonderfully sweet and there's a light florality to their flavour. Most 'cures' for flatulence are a load of nonsense (some even recommend *taking* inulin to cure flatulence, which has the exact opposite effect), so if they cause you pain I'm afraid you'll just have to avoid them. But of course, if you relish flatulistic adventures, these little powerhouses are your friend. Peel and roast alongside potatoes at Sunday lunch, then sit back and wait for the fun to start.

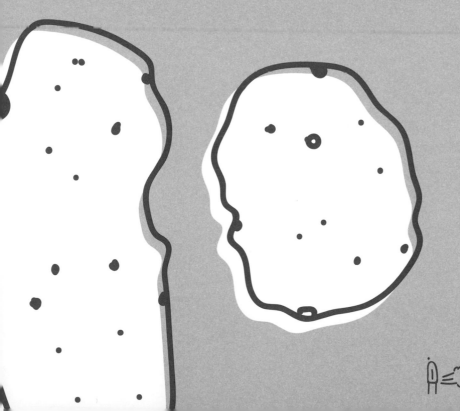

The world's fartiest recipes

If you have particularly adventurous guests, why not invite them round for a fart-fuelled dinner party?

ROCKET FUEL ROASTED JERUSALEM ARTICHOKES

This combines the notorious Jerusalem artichoke with roasted alliums to create as much inulin-fuelled wind as possible (from the artichokes) with a truly sulfurous stench (from the roasted aliums). It's so powerful that I am banned from cooking this in my house any more. This dish is a useful tool for measuring metabolic transit times, if you're interested. Just make a note at the time of eating, and again when any increase in your normal farting profile begins.

Serves 2 as a small meal or 4 as a side dish

750g/1lb 10oz Jerusalem artichokes
juice and zest of ½ lemon
4 small red onions
1 whole head of garlic
1 tbsp fresh thyme leaves
1 tbsp fresh rosemary leaves
2 tbsp extra virgin olive oil
pinch of salt and pepper

2 tbsp pine nuts
6 rashers streaky bacon, diced
1 small handful of parsley

1. Preheat your oven to 180°C/350°F/Gas Mark 4. If your Jerusalem artichokes are neat and clean (there are lots of different varieties), scrub them with a nail brush, chop into large chunks and toss in the lemon juice to stop them turning brown. Otherwise, peel them, then chop into large chunks and toss them in the lemon juice.
2. Peel and quarter the onions.
3. Chop the garlic bulb in half.
4. Throw all the veg into a roasting tray along with the lemon zest, herbs, olive oil and seasoning (but not the parsley).
5. Roast for 40 minutes. Meanwhile, toast your pine nuts and bacon separately in a small frying pan, then set aside.
6. After the 40 minutes' roasting check the artichokes. If they are nicely browned and crispy you're done. If not, give them another 10 minutes and check again. Test for seasoning.
7. Scatter parsley, bacon and pine nuts on the artichokes and serve.

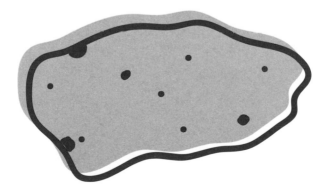

BEET THE TOILET

The beautiful thing about beetroot is the way it stains your poo red due to a red dye called betacyanin, most of which stays intact through the entire metabolic process, depending on your stomach acidity. Beetroot is packed with dietary fibre for high volume farts, and this recipe adds allium for that sulfur-containing stink.

500g/1lb 2oz fresh beetroot, scrubbed and cut into big wedges
6–8 garlic cloves, peeled but left whole
6 tbsp olive oil
1 tsp fresh thyme leaves
2 large red onions, peeled and sliced into thin rounds
salt and pepper
2 tbsp red wine vinegar
1½ tbsp brown sugar
4 tbsp crème fraîche

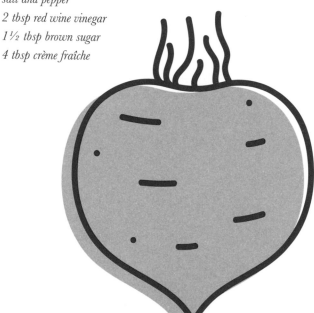

1. Preheat the oven to 180°C/350°F/Gas Mark 4. Put the beetroot, garlic cloves, 3 tablespoons of the olive oil and the thyme leaves into a roasting tray and toss to coat. Put the tray into the oven and roast for about 45 minutes until crispy around the edges.
2. While the beets are roasting, add the remaining olive oil to a large frying pan over a low heat. Add the onions and a good pinch of salt and cook for about 20 minutes, stirring constantly.
3. Once the onions are soft and browned, add the vinegar and sugar and stir until evaporated and sticky. Set aside.
4. When the beets are done, toss them with the caramelised onions and serve with the crème fraîche spooned on top.

PHYSIOLOGICAL SURPRISE!
EASY DOUBLE-STENCH, URINE-DYEING SALAD

A great one to try out on unsuspecting friends. Asparagus contains some extraordinary compounds that break down into methanethiol and dimethyl sulfide to make your wee smell really strongly of rotten cabbage, so combine this with the phenomenal gas production capacity of Jerusalem artichokes and the faeces-dyeing ability of beetroot and you've got yourself a festival of physiological quirks.

300g/10½oz beetroot
1 bunch of asparagus, cut into 5cm/2in long pieces
300g/10½oz Jerusalem artichokes, peeled and chopped into
  £1 coin-sized slices

1 small handful of fresh basil leaves
For dressing:
juice of ½ lemon
1 tsp Dijon or wholegrain mustard
2 tbsp really good extra virgin olive oil
1 tsp honey
salt and pepper

1. Boil the beetroot for 40 minutes on a low simmer, then drain and leave to cool. When they are cool enough to touch, skin them using a flat knife and just slide the skins off. It should be very easy. Chop them into wedges and set aside.
2. Meanwhile, bring a large pan of water to the boil, then add the Jerusalem artichokes, followed by the asparagus. Blanch them in the water for no more than 2 minutes then drain. Put all the veg together in a bowl.
3. Combine all the dressing ingredients in a jar, seal and shake until they've mixed well. Pour over the veg and toss thoroughly. Transfer to a serving bowl and scatter over the basil leaves.

ULTIMATE FART CLUB

Have you seen Mel Brooks' *Blazing Saddles*? Nuff said.

500g/1lb 2oz good sausages
olive oil, for frying
6 rashers streaky bacon
2 x 400g/14oz cans cooked haricot or butter beans, drained

400ml/14fl oz passata
salt and pepper

Fry your sausages in a little olive oil for about 10–15 minutes until nicely browned, then set aside. Add the bacon and fry until crispy, then put the sausages back in and add the beans and passata. Simmer uncovered for about 10–15 minutes until the tomato sauce has begun to thicken a little. Season, serve and sit back.

DUCK FART COCKTAIL

There's no flatulent outcome from this Alaskan invention, it's just a great cocktail with a good name.

Take a large shot glass or sherry glass and slowly pour in:

1 measure Kahlúa
1 measure Baileys Irish Cream
½ measure Canadian Whisky

That's it.

A young man is visiting his girlfriend's parents for the first time. Understandably nervous as he sits with them at the dinner table, he needs to let out some gas.

Luckily for him, the family dog, Buck, comes over begging to be petted, so he slips out a fart and waves his hand in the air a couple of times as he tells the overly friendly dog,

'Go away, Buck, that's enough.'

After this happens a couple more times, the father adds,

'Do what he says, Buck, before he shits on you.'

Who Cut the Cheese?
By Jim Dawson
(Ten Speed Press, 1998)

Are farts bad?

Broadly speaking, no. On one hand the hydrogen in your farts could be highly explosive; and if you combine hydrogen and oxygen together you get a powerful *hydroxy* explosion, which is terrifying, as my neighbours can affirm (methane explosions are mild in comparison). However, the concentrations of hydrogen and oxygen in your farts are low, and they are usually dominated by inert nitrogen, so you don't pose too much of a risk. Farting is a perfectly natural, normal function of digestion and should simply make you feel proud that you've eaten a healthy amount of fibre.

So, when are farts bad?

1. When you are a cow. As well as your farts, your cud-chewing, fermentation-based 'rumination' digestive system produces *huge* amounts of carbon dioxide and also methane in your belches, which is a potent greenhouse gas. Naughty cow.
2. When you are in pain because of 'trapped wind'. We all get this from time to time, but if you have persistent pain, bloating, really excessive flatulence or think that your farts are revoltingly smelly, it's probably best to see a doctor in case it's a symptom of a different problem. Off you pop and don't be shy.
3. When you're so embarrassed by either the sound or the smell of your farts that your quality of life suffers (because you stop socialising) or your diet suffers (because you avoid fibrous foods).

Although this book is a celebration of flatus, it would be obtuse to ignore the general perception that farting is embarrassing and rude. It's a bizarrely judgemental, prudish state of affairs, and it would be better if we all grew up a bit and offered the fart the airing and respect it deserves, but that's unlikely to happen any time soon. If you're determined to curb your wind, see page 108.

Surely fart gases can't be good for you, can they?

Every single component of a fart *could* kill you at a high enough dosage, but then so could a high enough dosage of water, perfume, carrot juice or hamsters. The basic principle of toxicology (the study of poisons) is that it all depends on the dosage (how much of them you consume). Water intoxication can kill – if you drink too much water too quickly the balance of electrolytes in your body is disrupted and cells in your brain can swell, causing cerebral edema and interference with your central nervous system. This goes back to a chap called Paracelsus (1493–1541) who said 'Everything is a poison, there is a poison in everything. Only the dose makes a thing not a poison'.

So yes, every single one of the components of your farts *could* kill you if you consumed them to the exclusion of anything else for long enough. But that would be pretty hard to do.

On the other hand, a study at the University of Exeter has investigated the potential benefits of hydrogen sulfide (the gas that gives our farts their rotten egg smell). Although this gas is dangerous in high concentrations, minute quantities have proven to be protective of cell mitochondria, which supply energy to a cell and can be damaged by diseases. It's unknown whether the lungs can absorb the hydrogen sulfide from a whiffed fart in practice, but it's fascinating to know that a gas that was hitherto considered dangerous could potentially be used to reverse disease in cells.

It would be obtuse to avoid the fact that although we love the smell of our own farts, people generally hate smelling someone else's. Ownership is everything in the world of flatus, and we don't like an uninvited whiff of someone else's bowels – partly because we are hardwired to avoid poo as it contains bacteria that can make us ill (it's not a waste product for nothing). Shame, really.

What are the bacteria in our guts, and are they good or bad?

You have around 200g/7oz of bacteria in your colon at any one time, made up of 100 trillion bacteria of over 700 different species, alongside a host of other microbes such as achaea, fungi and protozoa – all busy wriggling around inside you throughout the day and night. Numbers on this scale quickly become meaningless but bear in mind that there are thought to be only 37 trillion human cells in the entire body, and you'll begin to realise the impact this world of tiny creatures: you have more little alien creatures inside you than you have cells of yourself. We are only beginning to appreciate the importance of this world on our health and happiness, which is why all those bacteria are now seen as one collective functioning group, and have been given names such as the *biome*, *gut flora* and the 'forgotten organ'. It may even control progression of neurodegenerative diseases like Parkinson's and Alzheimer's.

Bacteria may cause horrible diseases, but that doesn't mean that they are all bad – in fact they can be very good indeed. It's known as a symbiotic relationship: we host our bacteria and give them a cosy colon to live in as they feed and reproduce, and in return they produce essential vitamins and minerals as well as all the gas we fart, and they break down our food for us.

Are bacteria good or bad?

I'd love to give you a definitive answer on this, but *it's just not that simple*. Different bacteria co-exist in our gut and they have such complex relationships with our bodies and each other that we barely understand them yet. Some of them are perfectly safe in our guts but potentially dangerous outside it (salmonella, *Staphylococcus aureus* and *Clostridium tetani* all cause serious disease); you have to remember that every substance on earth is toxic at *some* dosage. Some microbes are actively beneficial to us and some don't seem to be. Some ferment dietary fibre into acetic and butyric acids, synthesise vitamins B and K and metabolise bile acids and produce hormone-like compounds, which would seem to make them beneficial. Although not yet fully understood, it's thought that the gut biome plays an important role in lots of areas of medicine, from mental health to inflammatory and autoimmune issues.

Although there are between 300 and 1000 bacteria species in your gut, 99% of the total population come from only 30–40 species. The main ones belong to the following groups, in roughly descending order of incidence:

Faecalbacterium

Bacteroides (*there are several different species of this in our gut flora and about 30% of all gut bacteria belong to this genus. Normally mutualistic (good), with species found especially numerous in people who eat plenty of protein and animal fats*)

Escherichia (*such as* Escherichia coli)

Eubacterium (*such as* Enterococcus faecalis)

Enterobacter

Klebsiella

Bifidobacterium

Staphylococcaceae

Lactobacillus

Clostridium

Proteus

Pseudomonas

Salmonella

Prevotella *(found in people who eat lots of fibrous carbs)*

There are also these genera of fungi:

Penicillium

Candida

Saccharomyces

Rhodotorula (often found in people with IBS)

Pleospora

Aspergillus

Sclerotina

The joy of faecal matter

Poo *(faecal matter if you're being formal)*

Going for a poo is commonly known as 'laying some cable', 'baking a brownie' or 'taking a crap' (the latter was thought to originate from Thomas Crapper & Co toilets in the late nineteenth century, but in fact it was in use much earlier). In rhyming slang you could 'take an Eartha' (Kitt), a 'William' (Pitt) or a Brad (Pitt).

If faecal matter (poo) is the mother of the fart, then urine is its brother, sweat is its sister and nasal mucus is its uncle. Scabs, ear wax, saliva, vomit and belly-button crust are all quirky second cousins you only meet at weddings and funerals... and they prove that biological/anthropological metaphors only work up to a point.

Poo is metabolic waste matter, which means it's the stuff left over from the digestive process, as well as a range of other products that the body would like to get rid of. Men's bowel habits differ from women's, with men averaging 9.2 stools per week but women only 6.7; 40% of men lay cable once a day, compared to 33% of women. While 7% of men bake a brownie two or three times a day (your humble author thought that his regular three times a day was normal until very recently) but only 4% of women do the same. And 1% of women drop an Eartha once a week or less. Ouch. The most popular time to take a Brad is early morning, with men going earlier than women.

What's in it?

Poo contains a combination of foods that weren't absorbed in the small intestine (so were instead rotted down by bacteria in the large intestine), alongside excess or dead bacteria, waste products from the metabolic process, and dead gut cells. And it's all covered in a handy coating of mucus to help ease its passage out of your bum.

- 30% insoluble dietary fibre (food that cannot be digested, such as sorbitol, cellulose, inulin)
- 30% bacteria, both dead and alive (these are constantly replaced)
- 10–20% inorganic matter, such as calcium phosphate
- 10–20% fats, such as cholesterol
- 2–3% proteins
- Dead cells from the gut lining
- Bilirubin, a yellow substance produced by the breakdown of old red blood cells
- Dead white blood cells (leukocytes)

Everyone's poo is different and will change on a daily basis depending on what you've eaten, how well your digestive system is working and how healthy you are. Although poo is a perfectly normal part of human digestion, our obsession with safe sanitation is well-founded. Those bacteria and other pathogens are not meant to get into our mouths or stomach, where they can cause terrible problems. Wash your hands!

The Bristol Stool Form Scale

The Bristol Stool Scale is one of Bristol's many great gifts to mankind alongside suspension bridges, tarmac and Concorde (and just to be absolutely crystal clear, 'stool' means a poo). The scale was developed at Bristol Royal Infirmary in 1997 and it's also, alongside the BBC's official list of most offensive words *in order of offensiveness*, one of those fantastic documents that you can't believe anyone ever felt the need to write down, but on balance you're glad they did. It's the descriptions that really get you; for instance, 'like a smooth, soft sausage or snake'. That's normal, apparently. In case you're interested, my personal favourite is Type 3 – 'like a sausage, but with cracks in its surface'.

Type 1	*Separate hard lumps, like nuts (hard to pass)*
Type 2	*Sausage-shaped but lumpy*
Type 3	*Like a sausage, but with cracks in its surface*
Type 4	*Like a sausage or snake, smooth and soft*
Type 5	*Soft blobs with clear-cut edges (passed easily)*
Type 6	*Fluffy pieces with ragged edges, a mushy stool*
Type 7	*Watery, no solid pieces; entirely liquid*

You might wonder why we need a Stool Scale – how many people have weird poo, for crying out loud? Well, you might be surprised to learn that only 61% of men's stools are classified as normal, and only 56% of women's. There are a LOT of wonky stools out there, and what's normal to you may actually be pretty unusual. Until someone actually bothered to create a scale, you wouldn't have known.

Faecal transplants

Brace yourselves: this is going to be REVOLTING. Some people suffer so badly from devastating but little understood digestive disorders such as irritable bowel syndrome (IBS) that they resort to drastic measures. One such is the faecal transplant, which I highly recommend that you *never* try. The idea is that IBS may be due to an unbalanced population of bacteria in the gut (also known as your 'biome'), perhaps one or more strains of bacteria have multiplied more than others, and that this may be the cause of the problem. So desperate are some people to resolve this that they… there's no way to describe this pleasantly… take a fresh poo from someone healthy and squirt it up their bum. The hope is that the healthy person's poo will have a better balance of bacteria and that these will grow and multiply in the right balance and cure the problem.

At first glance, it makes sense but – and it's a whopping great 'BUT' – the benefits or drawbacks of changing your gut bacteria are little understood and there are reports of devastating side-effects. At least one operation seems to have ended with weight gain and others affecting mental health. It's performed in some clinics in the US and there are lots of reports of DIY home transplants. Don't do it. It's risky, it's not understood properly and it could cause worse problems.

How does food get pushed through our bodies?

Food and drink is pushed all the way through your body from mouth to anus by a series of wave-like contractions and relaxations of your gastrointestinal tract – these are called 'peristalsis'. You may not think that this is much of an achievement but bear in mind that your digestive tract is up to 9m/29½ft long. (Actually, it's 9m/29½ft at autopsy when it's completely relaxed, but is thought to be a bit shorter in the living body as it's usually in a state of tension.) Imagine a 9m/29½ft long tube of toothpaste that squeezes the food out on its own, or a vast earthworm (earthworms do use a spookily similar system to get around) and you're sort of there. Most of your digestive tract is surrounded by circular muscles that work together to create this motion and you have no direct control over it.

As soon as you've chewed a mouthful of food, gastroenterologists refer to it as a *bolus*, and the first peristaltic action takes place in your oesophagus when this bolus is forced down towards your stomach. Nerves around the oesophagus sense it moving down, and the muscles relax in front of its path to let it through and contract just behind it to push it along. Very clever, all choreographed by your nervous system without your having to think about it. My researcher friends, Alex Menys and Heather Fitzke took me for an MRI scan at University College Hospital London and created a video of peristalsis happening inside me. It's an extraordinary video, and you can watch it on the YouTube channel GastronautTV.

Once inside the stomach, those pesky gastroenterologists change its name again to *chyme*. And in case you've just vomited up your food, you may be interested to know that throwing up is *not* a peristaltic action, but a simple contraction of your abdominal muscles.

Once the stomach has processed your mouthful, the chyme is pushed through your *pylorus* (which looks like a sphincter but isn't one) and into your small intestine – small in width (about the width of your middle finger) but very long in length at around 6m/19½ft. Here, peristalsis slows down to a snail's pace (unless you have bad diarrhoea and your body is trying to get rid of bad food very quickly) because part of the function is to mix the chyme together with the enzymes that are breaking it down in to its components, then to absorb the resulting molecules through the intestinal walls and into the bloodstream for use by your body.

Once the small intestine has finished with the chyme, it enters the large intestine (1.5m/5ft long and 6–7cm/2½–2¾in wide) for bacteria to break down anything useful that remains, and for water to be absorbed into the bloodstream. Peristaltic action happens here, although the main transport comes from *mass movements*. This occurs a handful of times a day, triggered when you eat a meal. These push the chyme towards the rectum, which is basically a holding chamber that stores it until you're ready to go to the toilet.

Once inside the rectum the chyme has become faecal matter. The rectal walls expand as the poo enters, and once there's enough pressure, stretch receptors are activated, sending messages to your nervous system that you need to go to the toilet. If you don't go to the loo the faeces can be pushed back into the colon. This means that more water than usual is extracted, leading to drier, harder faeces and sometimes constipation. So if you've gotta go, you've gotta go.

If you're good to go, you sit on the loo (or better, squat), and the pressure builds up in the rectum until the faeces enter the anal canal – the last stretch before freedom. The rectum contracts as the faeces head south, then the last set of peristaltic waves push them out of the rectum, through the anus. The last piece of fascinating indignity is that both the internal and external sphincters pull the anus up and over the faeces as they make their happy way out.

Now wash your hands.

What would happen if we didn't fart?

Kaboom!

Actually, it's a little simplistic to say that if we didn't fart we'd explode – in fact it all gets a whole lot nastier than that. If you're really determined and able to hold your farts in, here are the likely effects – nicest first, worst last.

1. **Pain** Firstly you'll start to feel discomfort and a sense of bloating – and this will slowly become pain. Intestinal pain is a sign that something is not going as it should in your digestive system, so it would be crazy to continue, but if you did, you'd probably move swiftly on to indigestion and heartburn. And that's just the start.

2. **Farty breath** If they hang around your guts for too long these gases can also end up being reabsorbed into the bloodstream and excreted in the breath. That's not nice.

3. **Farty burps** If you persevere with holding your farts in, you could end up burping your farts. This is called reflux, and normally it's due to the contents of your stomach going upwards instead of downwards – such as when you vomit. But if you hold your farts in and they've got nowhere else to go, they can travel back up your intestines, resulting in a smelly, fart-flavoured, acidy burp. This chaotic return flow of air can happen to people who have IBS, as wind seems to pass more slowly through their guts.

4. **Gut busts** Diverticulosis is a common disease and holding your farts in is thought to be a factor in its development. If gas builds up it can create pockets in the intestinal walls – and if they become inflamed they can lead to perforated diverticulitis. And if these become septic without being diagnosed quickly you can die of sepsis. Not nice.

Why are some farts hotter than others?

We've all been there: sitting at your desk with half an eye on that boy/girl you fancy a bit – but largely minding your own business – when you feel a pressure in your rectum that says a fart is waiting at the gate. You take a glance at the boy/girl and think, 'Well, it's just a little one', so you lift your left buttock ever so slightly, relax that sphincter and do your best to control the release so that it won't be too loud. As it eases out silently you think to yourself, 'Dammit, fartmeister, you're good.' Then your pride turns to dismay as you feel an unexpected heat warming your pants. Oh God, no, you weren't expecting one of THOSE! Instinctively you know you've got a filthy, dirty, putrid bog-stinker on your hands and there's only one thing for it: blame someone else. So you glance around, tut loudly and put on your most innocent look. Nobody's fooled, of course, so you die inside and wonder why you ever left the house that morning.

The science of hot farts

So why are those silent-but-deadly farts so *hot* and stinky? It's all down to what happens when your bacteria break down dietary fibre in your gut. The process is called *metabolism* – a series of chemical changes in the cells of organic matter whereby fuel is converted for use by your body, but also some new components are built. When bacteria convert fuel (e.g. glycolysis when glucose is broken down into pyruvate) this destructive 'breaking down' process is called *catabolism*, whereby complex molecules are broken down into simpler ones, and in the process they can release lots of heat – hence hot farts. This transformation from chemical into thermal energy is called an *exothermic* reaction (one that lends to an increase in temperature). This happens because molecules like sugar store energy in their bonds. When they break down some of these molecules are fast-moving – i.e. they are hot.

So, hot stinky farts tend to come when the conditions just happen to be perfect for turbocharged metabolism of food: when there's loads of fuel available to your gut bacteria (i.e. you've eaten a great deal of dietary fibre), when your gut is super-populated with active bacteria either because it's been well-fed with fibre over a long period of time or possibly you've eaten lots of probiotics, and it's at optimal operating conditions such as perfect internal heat and levels of acidity. So both volume and smell should be at extremely high levels. Think yourself lucky!

About those infrared videos of people farting...

There are lots of videos on the internet that claim to be secret infrared camera footage of people farting as they go about their day. They're very funny, but we're pretty sure they're all fakes. We have an excellent infrared camera for work, and we've filmed many farts looking for that elusive cloud of joy, but it's impossible to spot them. Someone's just had a fun idea, and faked it with some fun graphics to make you gasp.

Chapter 03:
Fart Physics

"PARP!!..."

TRUMP!

TRUMP!

PFFT!

HORUMPH!.

"PUMP..."

Why do farts make that noise?

Welcome to the field of *dirty* physics – specifically anal acoustics (the study of arse-created mechanical waves that result in sound) and sphincter-specific fluid dynamics: the study of the movement of gases and liquids through the anus. There'll be a fair amount of discussion of anuses and sphincters here, so brace yourselves. (I've studied more diagrams of anuses than anyone without a medical degree ever should.)

Sound comes from vibrations that create lots of pressure waves. Humans can only hear those pressure waves if they happen a certain number of times every second (the frequency) – specifically between 20 times per second for very low bass sounds (20Hz), to 20,000 times per second for very high treble sounds (20KHz). So, for a fart to be heard, it must come from something vibrating between those numbers. That something is your anus – specifically the external opening of the rectum, tightly controlled by two sets of circular muscles known as the inner sphincter and the outer sphincter.

As the fart gases build up in your rectum – the storage tank for gas and poo – the pressure also builds, which you can feel as a need to fart or poo because a clever set of tiny mechanoreceptors send messages to your brain saying 'mind yer backs, big load coming down'. These sensations are clever, which is why you can usually tell the difference between a fart and a poo. When you decide to relax your outer sphincter (you have no control over the inner sphincter, but you can control the outer one), the pressurised gas is able to push open a small hole through your anus.

But why does your anus *vibrate* when you release a fart, causing that all-important raspberry sound? Well, it's all about pressure and friction. What happens is that the sphincter opens just a crack to let the fart out, but as soon as the gas is moving it sucks the anal sphincter back together as the fart flows through it partly because a faster flow creates lower pressure, partly because it curves around the edge of the sphincter as it goes, and partly because as soon as the hole opens, the pressure drops ever-so-slightly in your rectum. This closes the hole momentarily, but almost as soon as it closes, the pressure builds up a little more, pushing the hole open again, lowering the pressure to shut it again, and on and on with a fast opening and closing action. If that opening and closing happens at least 20 times a second, bingo, you·create a series of pressure waves within the audible range and you have a fart! That's where fluid dynamics comes in*. Weird, isn't it?

* One for the physicists amongst you. A key principle of fluid dynamics is known as Bernoulli's principle: an increase in the speed of a fluid occurs simultaneously with a decrease in pressure – though, as any engineer worth their salt knows, this only works within a streamline (the field lines of a fluid flow). The trouble is, defining where the streamline actually is once your fart passes your internal anal sphincter, reaches your external anal sphincter and escapes into the open, gets a bit complicated (see pages 86–7).

So, a fart is a battle between the high pressure inside the rectum, with the lower pressure it creates as soon as the fart is moving out through your anus.

You can also change the sound of your farts by tightening or loosening your sphincter as the fart flows out – the tighter you squeeze, the higher the note should be as you increase the pressure of the gas in your rectum – and the faster it vibrates due to the tighter sphincter and smaller hole.

Obviously you are at risk of either letting some poo out if you loosen the sphincter too much (commonly known as a *shart*) or, worse, stopping your fart altogether. Naughty sphincter.

You can use this sphincter control for entertainment purposes (see page 129) or to save your relationship with loved ones. If, for instance, you are awakened by the sensation of a pre-dawn whopper about to blow but you'd rather not wake your bed partner, just get yourself ready to drop the air sausage, and then, as the gas exits, hold your bum cheeks wide apart so that the sphincter is held firmly open. The sphincter will be unable to vibrate to produce that familiar rasping sound, and instead your rectal gas will escape with a mere curtain swish of wasted potential, leaving your partner blissfully unaware (depending on the aromatic toxicity, obvs). Obviously watch out for follow-through, but most farters worth their salt should be able to pull it off. Then again, if it's 7 a.m. and about time your partner got their sorry ass out of bed, pull your buttocks and sphincter as tight as a bow and let rip with all you're worth. Morning!

Bernoulli's principle or Coandă effect

Daniel Bernoulli (1700–82) was a Swiss physicist and mathematician born into a family of bitter, jealous, scheming scientists. His particular genius was in applying maths to mechanics and he's remembered in *Bernoulli's principle*, which describes the conservation of energy in fluid dynamics. Sounds dull, but its application is fascinating and it's the basic principle of how carburettors work, and *possibly* how farts move at different pressures and speeds in different situations.

The principle is this: within a flow of fluid, points of higher fluid speed will have less pressure than points of slower fluid speed.

A classic science demo that *purports* to show this in action is to levitate a ping-pong ball in a flow of air from a hair dryer or, my personal favourite, levitating a beach ball with a leaf blower. There's a fair amount of debate among physicists as to whether or not the effect relates to the Bernoulli principle or the Coandă effect, but analysis of air flowing around a ball using a beautiful visualisation technique called Schlieren optics clearly shows that as the ball moves away from the central flow of air and moves to one side, the air flow on the outside of the jet is slowed down compared with the faster-flowing middle of the jet and the ball is being pulled – and pushed – back to the middle. Bernoulli's principle applies to a closed system, but it's thought to apply here, too. Similarly as your fart rushes out of your sphincter it creates low pressure, sucking the sphincter back in with the help of the momentary drop in pressure after it has opened to let the fart out.

Romanian engineer Henri Coandă (1886–1972) was a rubbish soldier but an awesome engineer who claimed to have created the first jet with his Coandă-1910 (although his colleagues and contemporaries weren't convinced his was the first). His aerodynamics work recognised that a fluid jet tends to stay attached to a convex surface and develops a region of low pressure. This became known as the Coandă effect, and is another, possibly better, basis on which to understand how the levitating beach ball stays in the middle of the flow of air. If the ball drifts out of the main jet of air, the air on the convex surface in the middle of the jet flows faster (and hence is at a lower pressure) than that in the less speedy, more turbulent (and higher pressure) outer part of the jet. This pulls (and pushes) the ball back into the middle. It's pretty complex stuff, but the same pressure changes are likely to happen around the edges of your sphincter.

How to build a leaf blower-powered mega sphincter

Simple. First you need a huge sphincter.

1. Either borrow a rubber glove from under the sink or buy a 1m/3¼ft diameter balloon (the latter are particularly awesome). If it's a glove, chop the fingers and thumb off to leave a single rubbery sleeve. If it's a balloon, cut the balloon in half crossways (i.e. not through the neck of the balloon – the other way).

2. Get your hands on a leaf blower. Beware, these things can be dangerous – never, ever, EVER point it at your face. If just one little bit of grit got sucked into it, that could destroy your eye. Don't do it.

3. Pull the thinnest end of your glove or balloon over the end of the leaf blower and secure it with strong sticky tape.

4. Put earplugs in your ears, or put some headphones on, then get a friend to hold the leaf blower. Grab hold of the gloves/balloon at the loose end with two hands and pull it apart in readiness, standing to the side of it so that it's not pointing at you.

5. Now shout to your friend to turn it on. As the air comes blasting through the leaf blower, pull and relax the sphincter to adjust the sound from big flapping belter to screaming banshee horrorfart.

6. Happy days.

How does a sewage plant work?

The journey of your poo, pee and assorted toilet trash through a sewage plant has surprising similarities to the human digestive process – there's bacterial breakdown, some serious top-of-the-range Grade A gas creation and constant transit through the system that takes the place of peristalsis. The end product is clean water and the finest, sweetest fertiliser-poo for laying back on to the land (handily completing the *nitrogen cycle* essential to life). And if that wasn't fascinating enough already, sewage plants are grand-scale engineering projects complete with whopping great Archimedean screws, vast settling tanks and futuristic-looking biodigesters. Really, there's something for everyone – I don't know why people bother with Disneyland. I was taken around Wessex Water's Avonmouth sewage plant near Bristol by my friend Mohammed Saddiq and I liked it so much that I've been back several times.

The basic job of the sewage plant is to separate household sewage, industrial effluent and urban run-off water into its constituent parts: treated water, usable methane, fats, fertiliser sludge and unusable solids. They work on the principle that everything should be recycled if at all possible, with clean water returned to rivers or the sea and only the solids and larger rubbish (nappies, sanitary pads, rags and cotton buds) being sent to an energy recovery plant (nothing gets to landfill).

After you've flushed the loo, the assorted gunk travels out of your house into a network of pipes and sewers linked together by pumping stations to keep it all moving until it reaches the sewage plant. That's unless you've installed a nifty home sewage system, such as a septic tank or aerobic treatment system. Septic tanks are large containers usually sunk into the ground a little way away from your house, into which all your sewage flows to be further broken down anaerobically (without oxygen) by bacteria. They are considered to be only primary treatment systems as the sludge needs to be periodically vacuumed out, and the semi-processed effluent liquid flows straight into the ground itself.

On arrival at the sewage plant your wastewater usually goes straight into a vast Archimedian screw that pumps it upwards. There needs to be a constant gravitational flow of liquids through the system, so height is very important. Next comes *pre-treatment* where the sewage is simply pushed through a bar screen to filter out larger solids. Then comes *primary treatment* as the sewage is held in large, calm tanks so that the heavier solids will simply settle to the bottom, creating a thick sludge and the lighter solids, fats and grease will rise to the top in a murky scum. These are the large circular pools you'll have seen if you've ever driven past a sewage plant.

If there's been heavy rainfall and the sewage system becomes dangerously overloaded, some sewage plants will bypass the rest of the water treatment and store the water in large storm tanks, then pump the water back into the treatment process when it stops raining. Otherwise it is aerated and sent on for *secondary treatment* where dissolved and suspended biological matter is treated, broken down and removed using naturally occurring micro-organisms. Water can then be sent back into the system (although if it's released into sensitive ecosystems there can sometimes be a *tertiary treatment* stage of microfiltration of chemical disinfection).

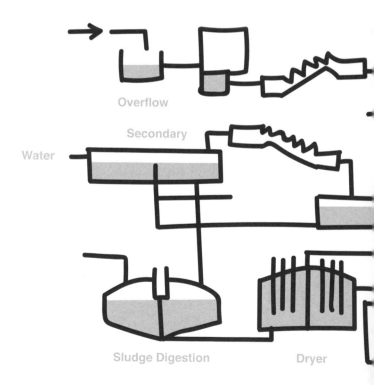

Overflow

Secondary

Water

Sludge Digestion   Dryer

The really fascinating part is the sludge digestion from both primary and secondary treatment. The solids that sink to the bottom of the tanks are removed and sent to an anaerobic sludge biodigester which works in a very similar way to your colon, where a varied population of bacteria breaks down the organic materials, producing a huge amount of methane (at Avonmouth they use it to power the plant and send biomethane to the gas grid) and more water. From here, the sludge goes into big spinning device called a centrifuge to remove water and produce a sludge-cake fertiliser. Large lorries collect the dewatered sludge for spreading on to land as fertiliser.

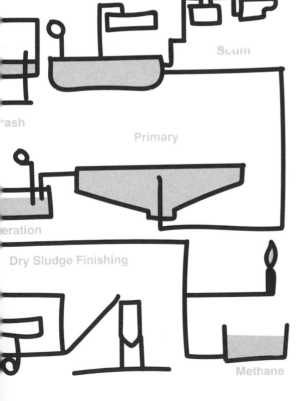

The fart bus

The official language of the canton of Ticino in Southern Switzerland is Swiss Italian, which means that the Regional Bus and Rail Company of Ticino is called *Ferrovie Autlinee Regionali Ticinesi SA*. Or FART for short, hence FART buses.

But this book isn't distracted by funny acronyms. Oh no. The fart bus we're interested in comes from a British company called GENeco – a green energy company that built a bus powered by biomethane generated from human and food waste instead of the standard diesel. I've taken a few rides on the bus from Bath to Bristol and it goes like a dream, and doesn't smell (although the awesome graphics on the side of the bus are designed to make the passengers look like they're sitting on the loo). The bus was a huge success, giving up to 97% reduction of emissions of dangerous particulates and an 80–90% reduction in nitrogen oxide emissions. Its carbon dioxide impact is also excellent, as you'd imagine.

GENeco also converted a VW Beetle to run on pure human waste methane, and it manages about 370km/230 miles on a full load of fuel, from several pressurised tanks in the boot. They let me drive it when we made one of my GastronautTV programmes at their base and it packs quite a poke. And no, it doesn't smell of farts.

The methane comes from a series of huge spherical biodigesters based at Wessex Water Avonmouth site near Bristol. They brew the methane from human sewage and recycled domestic food waste by feeding it a set of bacteria and sustaining the process at the correct acidity and temperature. The process is anaerobic, which means that the bacteria break it down in the absence of oxygen – and although not every human produces methane it's such a lovely project that I'm happy to call it a huge fart. Both the Fart Bus and the Bio Bug are proof-of-concept projects designed to show that the idea is workable, and the Bristol plant now produces a vast 56,000m³/1,977,621ft³ of biogas every day, which is used to directly replace natural gas and several bus companies are planning to use the technology.

How to build
a fart machine

*(or 'Anaerobic Digester' if you're currently
seeking departmental funding)*

This is a serious guide for teachers or home DIY/fart enthusiasts and involves live poo and liquids close to electrical items, so please take care and think about cross-contamination risks, and write yourself a full risk assessment. I got into trouble with my wife when I built a flavoursome anaerobic digester at home so I hope your family is onboard. It's a living, breathing organism so you have to check it constantly – get the acidity wrong and it dies. Develop a crust on it, and it begins to poison itself until it dies. Get the feed balance wrong and… you know the rest.

To build your fart machine, I've listed everything you need, which you can buy from a lab supplies company, but substitute anything you've got lying around. I recommend getting an immersion heater as the whole thing really needs to stay at a constant 37°C/98.6°F.

Oh, and once your machine is built, it needs a starter culture. You're best off pooing into a bucket and using that. I've read about people using cow pats but cows have a different digestive system so you may not get the bacteria you need to get a long-running machine. Here are a few pointers:

1. When you first get your machine up and running, you'll be amazed at how much gas he produces. Don't get too excited – it seems that a lot of the first few ventings are carbon dioxide, and not much methane. It needs to settle down – one attempt of mine worked straight away, but another misbehaved and took three weeks and a reboot! Of course, women's poo is most likely to be methane-producing, so if you're not a woman, you might have to ask one very nicely if they'd poo in a bucket for you.

How to build
a fart machine

2. Check pH and heat regularly, vent it and feed it regularly too.
    This is not a complete idiot's guide to making the machine –
    we're assuming you've got a bit of skill and common sense.
    Some items may need to be substituted or fiddled with to fit.

3. Teachers (and, in fact anyone doing this with other people
    involved) will need to write a risk assessment for this so you can
    get a handle on the issues at stake. Take particular care to think
    about hygiene and contamination issues.

The key lab kit you'll need:
   (Note: Check that everything you order is compatible (e.g. pipe
diameters and bungs). I've bought a lot of this kit from Timstar – the
lab supplies company I use – but feel free to get it from anywhere:

   Immersion heater
   Powerswitch power supply
   Stackable 4mm plug leads (red)
   Stackable 4mm plug leads (black)
   Ancillary equipment: LCD strip
   3 x 1litre filter flasks
   3 x 2-hole bung
   PVC tubing
   Disposable syringes

Red rubber tubing
Y-shape adaptors
T-shape adaptor
3 x Hoffman clips
Gas syringe
Retort base
Retort rod
Clamp, rubber covered
Bosshead
Gas jar
Beehive shelf
Pneumatic trough

Additional requirements:
Aquarium: 41 x 21cm/16 x 8¼in (L x W) minimum dimensions
1 litre/35fl oz/1¾ pints poo mixed with 500ml/17½fl oz lukewarm water
Feed: digestive biscuits, whole milk and icing sugar
Balloons
Optional
3 x Compact magnetic stirrers (with stirring bars)

How to build a fart machine

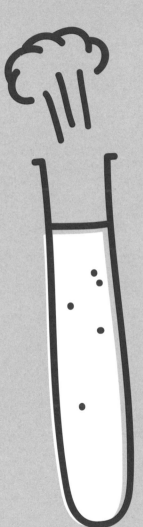

Set-up

1. I suggest that this is assembled on one or two trays so you can move it easily.

2. Assemble your fish tank with LCD thermostat and immersion heater connected to Powerswitch power supply (take precautions to isolate power supply from the water) and fill tank with water. Adjust Powerswitch supply to max. voltage of 12V to maintain water temperature of 37°C/98.6°F. Insulate it loosely so you can get access to it.

3. Cut 3 x 30cm/12in lengths of PVC tubing and insert each one into a separate 2-hole bung. Ensure that when the bung is placed in a 1 litre/35fl oz/1¾ pint filter flask, the tubing almost touches the bottom and at least 5cm/2in is visible from the top of the bung. These will be the 'draw tubes' to remove excess material.

4. Cut 3 x 10cm/4in lengths of PVC tubing and insert into the remaining holes of the bungs. Minimise tubing that's visible inside the flask after the bung is inserted, and leave at least 5cm/2in protruding from the top. These will be the 'feed tubes'.

5. Attach Hoffman clips to all visible tubing protruding from the top of the bungs (one per tube) and tighten to seal tubes.

6. Using red rubber tubing and Y-shaped adapters, connect outputs of each filter flask together so that they end in a single length of tubing. At the end of this attach a T-shaped adapter that leads into a 20cm/8in length of tubing and a 60cm+/24in+ length of tubing. This is the gas output system.

7. Fill the pneumatic trough with cold water and place next to the aquarium then place beehive shelf into trough. Have retort stand with claw and boss set up over beehive shelf to hold inverted gas jar filled with water.

8. Feed 60cm+/24in+ length of gas output tubing into gas jar so that it ends at the top. Attach either balloons or (when you're ready to demo some methane ignition) a gas syringe to the remaining 20cm/8in tubing of gas output system.

9. When gas is being produced by the system it will displace water in the gas tube. Providing gas output tubing is above water level, gas can be drawn off into the gas syringe for analysis. Ensure a Hoffman clip is closed at syringe output before removing syringe from the system.

How to build a fart machine

Start-up and maintenance

1. Place 500ml/17½fl oz of poo and water solution into each flask and firmly insert bungs. If using additional magnetic stirrers, place these underneath the aquarium below where each flask will sit, and ensure a stirring bar is placed into each flask before sealing. Lower all three flasks into the aquarium. These may initially need securing with tape otherwise they'll tend to float!

2. Leave poo for 2 days.

3. Mix 20g/¾oz icing sugar with 2 finely ground digestive biscuits. Dilute with whole milk until it is a thick, liquid consistency. This is the feed and can be made in bulk and stored in a fridge or freezer.

4. To feed, add some food to syringe, then push and connect it to feed tube of filter flask bung. Push-connect empty syringe to draw tube of same filter flask. Open Hoffman clips and insert feed into flask before drawing off same amount of digestate. Close clips before removing syringes and disposing of digestate.

5. After initial rest period, begin feeding gradually, increasing by 5ml/⅛fl oz per day per flask of sludge. Don't exceed 30ml/

1fl oz of feed per day and once this is reached split feeding into two 15ml/½fl oz feeds per day.

6. In both cases gas production rate will increase. Sludge systems will produce large gas volumes, but if production rate drops significantly from one day to the next stop feeding until gas production increases, then begin gradually increasing feeds again.

7. Bonne chance!

'Me wife's funny,' said one man. 'It was her birthday and she said she wanted to go to one of them fancy French restaurants. So we dressed up and off we went.'

'Yeah,' said his mate. 'What was it like?'

'Bloody expensive, though the grub wasn't too bad. But the portions were so small if you went outside and had a good fart you were hungry again.'

The History of Farting
by Dr Benjamin Bart
(Shelter Harbour Press, 2014)

Chapter 04:
Ask the Fart Doctor

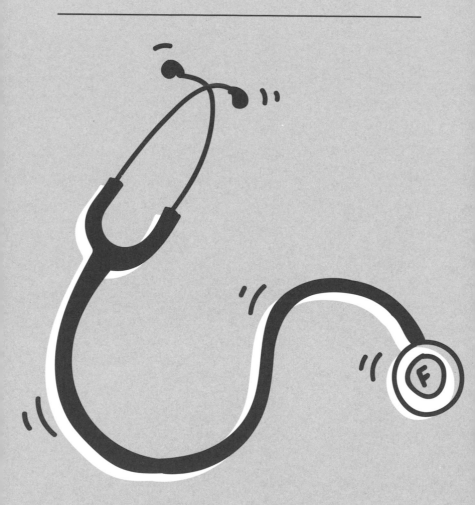

How to see what's happening in your guts?

have a beautiful body. The outside may be a bit iffy, but the insides are flipping *gorgeous*. I know this because I've eaten several cameras including a disposable PillCam (imagine a GoPro but smaller) to find out what's down there, and I've had several MRI scans too. I don't actually have any medical condition other than maniacal curiosity, but I can't seem to stop going down to take a look.

A gastroscopy (full title *oesophagogastroduodenoscopy*) involves inserting a camera down your throat, through your stomach and all the way to your duodenum, which is the first part of the small intestine. It's not the most pleasant procedure, but doctors love it because it's a useful diagnostic tool. Blood tests and bacterial cultures are great but there's nothing quite like getting down there and taking a good nosy around. If your digestion isn't up to scratch or your farts indicate that something's wrong, it won't be long before you'll meet a gastroenterologist with a nifty camera on a very long, bendy handle.

I've had three different gastroscopies: a PillCam, a small transnasal gastroscopy and a larger OGD (or EGD) oesophago-gastroduodenoscopy, and they were all fascinating, if a little undignified.

The PillCam is an ingenious little camera with integral light, all encapsulated in a pill about 2.5cm/1in long. It's the least invasive of the techniques and the most intriguing – it simply drifts through your digestive system as though it was a bolus of food, flashing its little light and filming as it goes. It has the advantage of drifting right through your duodenum, all the way through the entire small intestine and colon, and out through the rectum and anus. The downside is that you can't control it to take a look around at something specific and it may not be facing the right way as it passes by an ulcer or other abnormality. Incidentally, I tried to spot it sparkling in my poo over the next day or two but couldn't. Good job really as I would've been tempted to fish it out and pop it through again.

I also had a small transnasal endoscopy using a little camera (with a light) on a very small probe that gets pushed straight down your nose, through your oesophagus and onwards. I had mine while sitting upright in a chair, and the upside is that you can chat throughout the procedure (should you want to). Having a camera stuck up your nose is a bit uncomfortable but it's not painful, and the gastroenterologist can take a good look around as the camera can be controlled quite well. The downside is that using a camera small enough to fit up your nose means that it has relatively low resolution, so the pictures aren't very clear.

The full oesophagogastroduodenoscopy is my favourite internal inspection procedure and uses the big daddy of endoscopes: a high-resolution lipstick-sized camera on a fully controllable arm that comes with all sorts of special functions including lights, a water-squirting tool for flushing muck out of your guts, compressed air for expanding the guts so that the camera can inspect exactly what it needs to, and it has an unrivalled level of motion control, rotating 180 degrees to investigate every nook and cranny of your innards, of which there are many. A mouthguard is inserted between your teeth to stop you biting down on the camera as it goes down. It's all controlled with a series of triggers and buttons, and an experienced gastroenterologist can rival the best feature film camera operator. The drawback is that it's very uncomfortable, first as the camera is pushed down the back of your throat and then as you feel it moving around in your guts. It's also undignified because you're constantly gagging as your body tries to vomit up the camera and burp out the air that's being pumped in. You get a sore throat despite the banana-flavoured local anaesthetic that's sprayed down there beforehand. However, all these are minor complaints when you consider the stunningly high resolution of the video and the way it can be controlled to inspect every millimetre of your guts. Oh, and it saves you needing invasive surgery.

If you need a gastroscopy you may well be terrified of the procedure itself but take my word for it: although it's uncomfortable and strange, it's not actually painful, and you'll probably be surprised at how quickly it's all over. I had my oesophagogastroduodenoscopy courtesy of my lovely gastroenterology friends Phil Woodland and Heather Fitzke (and thanks to the Royal London Hospital). If you're interested it's all online at the YouTube channel GastronautTV.

How can I stop farting so much?

If you *really* feel that you fart too much there are a few ways to cut your output, but proceed at your peril – curbing guff production isn't always a good idea.

What is 'too much'?

The most common medical treatment for people who complain of farting too much is to convince them that they don't. A good digestive system produces a healthy 0.5–2.5 litres/17½fl oz–4¼ pints of gas every day. Most people don't like the amount that they fart, but that doesn't mean it's bad for them.

Change your diet

This is the biggest change you can make, but it's also the most dangerous. You can cut down on the amount of dietary fibre you eat, but on average we only eat only 18g/⅔oz dietary fibre per day in the UK, whereas the recommended consumption is 30g/1oz. I won't wag fingers, but bear in mind that meat, fish and dairy taste great but don't contain fibre. Fibre can help prevent heart disease, diabetes, weight gain and cancer and improve digestive health. Stop eating fibre and you're likely to be constipated. If you *promise* you'll be careful and consult your doctor, you might be able to cut your wind down by:

- Cutting down a little on beans (they usually contain complex sugars called oligosaccharides, especially raffinose, which is particularly good for gas production).
- Cutting down a little on vegetables full of dietary fibre such as Jerusalem artichokes, cabbage, cauliflower, onions, garlic.

People who are lactose intolerant lack the enzyme lactase so they can't break down the sugar lactose in their small intestine, leaving it to be broken down by gas-producing bacteria in the large intestine. If you have this problem cut down on cheese and other milk products but make sure you're getting enough calcium elsewhere.

Swallow less air*

Chew and drink more slowly, avoid sucking hard sweets or pen tops, stop smoking, and don't chew gum.

* This sounds strange, doesn't it? Well, there's a big difference between breathing and swallowing air (known as aerophagia). When you breathe, you draw air down through your windpipe (or trachea) at the front of the throat, and into your lungs. But when you eat, your epiglottis (a leaf-shaped flap of cartilage) covers the windpipe up so that your food is diverted into the smaller, more flexible pipe (the oesophagus), and on towards your stomach. So, although food isn't able to get to your lungs, it is possible for air and other gases to get to your stomach mixed up with the food or drink you consume, or when you swallow. Some gets burped back up but a fair amount carries on through the digestive system and either gets absorbed into the bloodstream or ends up being farted out. So, swallow less air, and you'll fart a bit less.

Stay grounded

Don't fly in a plane, become an astronaut or rock climber, because high altitude plays havoc with your guts. An Australian study showed that farting doubled 8–11 hours after a fast ascent up a mountain – most probably because lots of carbon dioxide is dissolved in the bloodstream and as you get higher, atmospheric pressure reduces, making that carbon dioxide diffuse, expand into the bowels and erupt! Commercial planes are only pressurised to the equivalent of 800–2,400m/6,000–8,000ft above sea level, so the same issue applies.

Shun sorbitol

Avoid chewing gum or diabetic sugar-free products (unless you're diabetic, obvs) as they often contain a sweetener called sorbitol, which isn't digested in the small intestine (hence it's a dietary fibre) but gets enthusiastically broken down by gas-producing bacteria.

Small plates

Eat smaller meals more often to slow down release from the stomach and encourage as much digestion as possible to happen in the small intestine, leaving less to the gas-producing large intestine.

Stop the pop

Drink fewer fizzy drinks. Some of that CO_2 will end up as flatus.

If you're worried about offending your neighbours on the plane, don't worry, those carbon dioxide-rich farts are likely to be a lot less smelly than normal.

Minted

Drink peppermint tea. Most studies on this relate to people suffering from irritable bowel syndrome (IBS), but it's known to calm the stomach muscles and relieve abdominal pain.

Ask the doctor

Ask your GP about the following:

- Alpha-galactosidase. Sounds like a sci-fi holiday destination but is actually an enzyme that helps to break down oligosaccharaides called glycolipids and glycoproteins.
- Probiotics. These can be a gastrointestinal can of worms (you mess with your gut flora at your peril), but some blends have been shown to have an effect.
- Antibiotics. We're on dangerous ground now, but one decent study of rifaximin has shown a clear effect of lowering gas production which diminishes, but remains significant, over time.
- Simethicone. This is an anti-foaming agent that breaks up bubbles in the gut and has been seen to be effective if you have acute diarrhoea.

Charcoal

Try taking activated charcoal tablets. Weird, huh? Activated charcoal has a highly developed microscopic internal pore structure (basically it's a sponge with a huge internal surface area) that encourages molecules stick to itself. Whether or not it really works when taken orally seems debatable (it's not particularly picky about which molecules it attracts), with at least one study saying it doesn't.

How can I make my farts less smelly?

1. Don't overcook brassicas (cabbages, broccoli, cauliflower) as the amount of hydrogen sulfide in them increases the longer you cook them.
2. Avoid beer (men only). One of the few decent studies on flatus identified a clear correlation between men's beer consumption and flatus aroma, though less so for women.
3. Avoid meat and high-protein vegetables. Sulfur-containing amino acids in protein break down and form those stinky sulfur-based compounds.
4. Avoid beans, especially soya and pinto beans.
5. Avoid garlic, onions and asafoetida.
6. Avoid fatty food.
7. Try Pepto-bismol, containing bismuth subsalicylate. It binds sulfide gases in the gut.
8. Sit on a bag of activated charcoal or buy some fart-reducing underpants. No, really.

Can you buy fart-filtering pants?

Not only can you buy them. They work. One of the few serious studies into fart smells was intended to evaluate whether or not charcoal-lined pads could absorb the sulfur-containing gases that make up the worst of our guffs. After a series of undignified tests and rectal tubework, Suarez, Sringfield and Levitt concluded that 'Sulfur-containing gases are the major, but not the only, malodorous components of human flatus. The charcoal-lined cushion effectively limits the escape of these sulfur-containing gases into the environment.'

The pants are dead clever. They are usually made of quite substantial airtight material with a special lining and good elasticated waist- and leg-bands so that no gases – good or bad – can get through without passing through a membrane of activated charcoal. This activated charcoal has an extraordinarily high level of microporosity – it's basically a solid sponge made up of millions of tiny pores. One single gram of activated charcoal has a surface area of over $3,000m^2/32,292ft^2$, which means it can adsorb a large amount of gas or liquid (adsorption is when atoms, molecules or ions stick to a surface). Activated charcoal is often used in air and water purification, and to treat poisoning and overdoses (it can soak up the poison in the digestive system), sewage treatment, decaffeination and in gas masks.

And if you thought that fart pants were just for people who've already said goodbye to dignity, take a look at www.myshreddies.com who make some very comely – some might say sexy – knickers, jeans and pyjamas that by all accounts actually seem to work.

How can I *increase* my wind?

If you're already eating foods rich in dietary fibre but still want more from your guts, fear not. Here are some more tips.

Swallow air

Around 25% of your farts are from swallowed gases and you can increase this simply by swallowing more. You'll burp a lot of it out but a fair amount should still make its way all the way through. Try gulping down your food very fast, chewing on various items like pen lids to increase your saliva production and thereby encourage swallowing. Gum chewing is useful too.

Boil your cabbage for a long time

This is about smell rather than volume. The amount of sulfur compounds in your cabbage increases the longer you cook it, making your farts much smellier. Between the five and seven minutes point of cooking they double. It makes cabbage taste grim, but that never stopped my granny.

Eat food and drink full of gases

The two ingredients that food manufacturers love adding to their products are water and air because they're cheap. Carbon dioxide-rich fizzy drinks are a great idea, but milk foams such as those floating

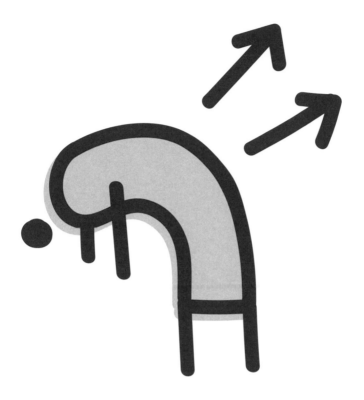

on a cappuccino or hot chocolate are also very stable and will sustain lots of air quite a long way down into your guts before collapsing and releasing their gas. There are also quite a few foods that have air bubbles cooked into them: meringues, whipped cream, honeycomb snacks, puffed snacks from corn, rice and potato starch, and rice cakes. Interestingly, whipped chocolate bars will often be made with nitrogen or carbon dioxide rather than air – but it all adds up.

Inhale air into your anus

With a little practice some people can 'inhale' air into their anus in order to be able to fart on demand. This is how the famous French flatulist Le Pétomane (Joseph Pujol) managed his copious flatulence. He was blessed not just with a quirky anatomy but also a quirky level of control over it. When he squeezed his nose tightly shut and contracted his diaphragm it made his abdomen expand and draw air in through his anus.

Here's a technique that used to work for me when I was a schoolboy but sadly no longer does. Perhaps it will work for you:

1. Lie down on your back with your legs up against the wall.
2. Pull your legs higher up (as high as you can go without feeling that your bum is clamping shut), and spread your legs a little at the same time (not too far or you'll get the same anti-farting response from your bum).
3. Relax your anus and block your throat so no air can be inhaled and try to pull up your diaphragm to create a sucking motion in your torso.
4. If that doesn't work, try sucking your chest in, spreading your bum cheeks apart or shifting your back around.
5. Persevere.

Release bowel stress

If you can feel pressure in your bowel already and know that there's a fart lurking inside you that just needs some encouragement, this is a technique sometimes suggested by gastroenterologists to help you pop one out. Lie down on the floor on your left side, bend your right leg at the knee to assume *Sims' position* (used for rectal examination) and roll on to your front, then back on to your left side – and repeat a few times. This will cause the bowels to shift around and put pressure on the descending colon (which is on your left side) where the fart ought to be residing, and hopefully this pressure will force the air downwards and generate a fart.

Avoid potatoes, bananas and wheat

Fart fans should avoid these because they can reduce smell (though not volume). Potatoes and bananas contain resistant starch while wheat contains fructans. The idea is that these foods are extremely fermentable, hence the bacteria in your colon gets so busy breaking down the carbohydrates that they never get to the protein, thereby producing less hydrogen sulfide. Boo.

Get infected with *giardiasis*

This is a warning, not an encouragement. *Giardia* is one of the most common parasitic diseases in the world and it makes you feel tired, sick and bloated and causes diarrhoea, vomiting, headaches, etc. Around 10% of infected people have no symptoms and it's spread by faeces contaminated with *Giardia lamblia* cysts. The thing is, people become temporarily lactose intolerant after infection, and if you consume dairy products this causes a surprising increase in farting. But it's not enough of an upside, all things considered.

Why are farts so embarrassing?

(Psychology of embarrassment)

Even the most hardened flatulist like myself can get embarrassed by an ill-tempered fart at an ill-timed moment. But *why*? These are natural functions essential to the functioning of our bodies – why is it that semi-voluntary bodily quirks like yawns and sneezes draw sympathy from our parents and partners but farts provoke disgust? Well, it's likely to be rooted in the ancient *miasma* theory of disease (see page 121) – that rotting smells and bad air are to blame for terrifying epidemics such as cholera. Hence a fart isn't just a display of base humanity because of the sound, it's tangibly *dangerous*, linked by bad rotting smells to disease, pain and death. The fact that it was comprehensively disproved after the 1850s is immaterial – the damage to the fart had been done. Couple that with the very real association between what comes out of your bottom and a host of horrible bacteria-related diseases and the deal's sealed. Farts are now socially unacceptable and it seems as though there's no going back.

Embarrassment theory is based on the idea that when an act is perceived as socially unacceptable (rather than morally wrong) it undermines the image of ourselves that we seek to project to others. So even though we may blame society for creating artificial constructs about what's right or wrong, the feeling of embarrassment is ours and not theirs – we have created it from *our* perception of what we want to project to others. Basically, we are to blame for our embarrassment.

The effects of embarrassment can be debilitating: blushing, sweating, defensiveness and nervousness, sometimes nervous laughter – these reactions can be intense and upsetting for some people. But *why*? What function does embarrassment serve? Well, there's an interesting theory that expression of embarrassment to others is an appeasement function and demonstrates that the farter knows they've transgressed but they know the behavioural rules and are apologetic for it and 'will prove worthy another time'. There are even studies that show people who are *not* prone to embarrassment are more likely to engage in antisocial behaviour.

Society has lots of abstract rules and expectations that seem to have no purpose, such as taking your hat off in polite company, standing up when a woman (or other superior being) walks into the room, not scooping up your peas with a fork in your left hand (yup, I can't do it) and not farting in public. Many obscure quirks like these are rooted in historic class delineations – you needed to apply yourself to a set of behavioural rules to be worthy of fitting in with upper societal levels, and as the middle classes began to aspire to gain access to those levels they would have to abide by these rules.

TRUMP!

Dr Seuss's beautiful story *The Sneetches* tears apart the concept of abstract social stratification. His Sneetches are divided into the upper classes who have stars on their chests and the lower classes who don't. When an entrepreneur turns up with a machine that puts stars on to the lower classes, they all pay him for stars, whereupon the upper classes are mortified and realise that they need their stars taken *off* to prove that they are different. The entrepreneur is happy to oblige, whereupon the lower classes also have them taken off and vice versa. This continues until everyone has forgotten who's supposed to be in which group and the entrepreneur is rich. The Sneetches realise that discrimination is ridiculous and they get along as friends. If only the rest of society's problems could be solved so simply.

Sociologists tend to be less negative, however, seeing manners as one step towards maintaining social order, and that stratifying society is one way of making centralised power and communal living bearable. I hate manners, but at the same time, I do understand that they may be useful. Basically, is it better to thump people to assert authority and work your way up to the top of your social group, or to set up an abstract system of manners and expectations of social behaviour, regulated by tools such as embarrassment, and thereby stop people from getting thumped? Not being a thumper myself, I prefer manners, but I still find them suffocating.

In the case of farts the behavioural rules that force us to keep them hidden do seem bizarre, causing physical and psychological discomfort, possibly leading to digestive problems and real pain Social stigma is hard to shift and I can't see it changing any time soon. But heck, we can still give it a try.

Why should you wash your hands?

Bad smells have had bad press since ancient times, and for hundreds of years they took the blame for terrible problems that could have been solved simply by washing one's hands. Up until 1880 the prevalent theory regarding the spread of disease was based on the concept of *miasma*, stinking 'bad air' from rotting organic matter – probably because cholera epidemics happened in places where undrained water smelled foul. In fact cholera was water-borne, not air- or fart-borne, but that was unknown until 1854, when the English physician John Snow made the connection. Even during the 1850s cholera outbreaks in London and Paris, many ignored Snow's discovery and worsened the problem.

Florence Nightingale, the English founder of modern nursing, was more switched on, and during the Crimean War in 1854 decided that one of the cheapest and most effective ways to reduce death rates was handwashing and basic sanitation. She was right.

But did we listen? Did we heck. The proportion of people who don't wash their hands after going to the toilet is shocking. A very large survey carried out across Europe in 2015 pointed a stinky finger at 62% of men and 40% of women who simply don't bother to wash

their hands at all. I am a trenchantly conscientious handwasher (although I'm pretty sure that when I was a schoolboy handwashing was for losers), but even I don't wash my hands for the full 15 seconds recommended by the NHS.

Perhaps you're thinking, 'Stop nannying me. My health is my responsibility.' Well, handwashing can reduce the incidence of diarrhoea by about 30% and bad hand hygiene is a factor in 50% of all food-borne illness (take a look at the Center for Disease Control and Prevention's handwashing 'Show Me the Science' page for links to lots of relevant studies. It's important, not just for yourself, but for your friends and family too). Handwashing with soap is regularly cited as the single most effective and inexpensive way to prevent diarrhoea, pneumonia and other acute respiratory infections. And bear in mind that pneumonia is the number one cause of death among children under five.

But more important is the risk that your dirty hands pose to someone *else* you might touch, even if they've been conscientious and washed their hands. You might be healthy and able to recover from disease without too much bother, but if your bacteria cross-contaminate from one surface to another and then on to someone's hands who then comes into contact with the old, ill or infirm, there could be tragic consequences. Come on, people. How hard can it be?

One of the fascinating things about bacteria is how easily they can spread by cross-contamination – simply by one hand touching some bacteria, then touching another surface, which is then touched by someone else who then touches their lips with their hand.

These bacteria can travel far and wide. In *Bodyhacking*, one of my big science stage shows, we coated a wig in UV-sensitive powder that glows under UV light and then attached it to a robot that wandered around the auditorium. We encouraged the audience to steal the wig and to throw it around so that lots of people ended up touching it, and they inevitably ended up touching their mouths, noses and each other. Later in the show we dropped all the lights except for some very powerful UV blacklights and pointed them at the audience to see how far the powder had spread. The results were surprising: the powder had spread far and wide, often far beyond the original contact point. It's a useful tool to explain cross-contamination but there is a downside: you can also spot who's been fiddling with whom in the dark, and this can lead to arguments.

Other physiological non-farty quirks

Hiccups

Hiccups are sudden and involuntary spasms of the diaphragm – and my youngest daughter suffers from them quite badly. Once the action is triggered, the hiccup happens a quarter of a second later. It's interesting that your brain doesn't control them – they are an automatic response from one part of your body to another using something called a reflex arc. The vexed question of 'what's the point in hiccups?' has bothered humankind ever since evolution stopped being a laughable idea.

Although only mammals have hiccups, one theory holds that they are an evolutionary remnant of creatures that live both in water and on land (amphibians). Tadpoles' respiration is similar to a hiccup (just one of the many reasons why it's rubbish being a frog – other than the dainty princess-snogging, which must be quite fun). Another theory is that it helps suckling infants to feed by encouraging trapped air to escape. There are no reliable cures for hiccups other than lots of sedatives or surgery on the phrenic nerve between the neck and diaphragm, which is fraught with complications as the phrenic nerve is important for breathing. That said, there's a paper in the *Journal of Internal Medicine* detailing how persistent rectal massage with a finger has been effective for some people, which brings up the question 'how much do you love your daughter?' to which the answer is 'not that much'.

Yawns

This thoroughly enjoyable reflex action falls into the category of 'very interesting but not worthy of much serious medical research because there's no money in it, squire'. It involves a deep inhalation of air, complete extension of the jaw, closing of the eyes, eardrum stretching and then exhalation of said air, and it's associated with tiredness, stress and boredom. Of all the theories as to why we yawn (and there are many) the one I like most is that it cools the brain down. I rather like the fact that we don't know what they are for, because if we found out, someone might invent a way to stop them and this would be a terrible shame, as I *love* yawning.

Sneezes

Sneezes are involuntary but although you do have some control over how they come out, that doesn't mean you should try. If you stop a big sneeze by blocking your nose and clamping your mouth shut you could rupture parts of your nasal passage, burst your eardrums, rupture blood vessels in your eyes, crack ribs, cause a brain aneurism or cause bubbles of air to enter the deep tissue and muscles of the chest. They've all happened.

You sneeze for a good reason – to get rid of foreign bodies that have irritated the mucus in your nose and to clear it out. These irritants trigger the release of histamines (hence antihistamines are offered as hay fever treatment), which trigger nerve cells in the nose to send tiny electric signals to your brain, kick-starting the simultaneous, involuntary, instantaneous response of your combined face, throat and chest.

The weirder quirks about this weird quirk are that some people are triggered to sneeze by eating a very large meal. Deaf people don't make much noise at all when they sneeze, and in the Philippines they say 'ha-ching' rather than 'ah-choo'.

Rumbling stomach

The medical term for a rumbling tummy is *borborygmi*, and the sound is made by fluids and air bubbles passing through your intestines. These sounds are sometimes produced by the stomach sending messages to the brain two hours after the stomach has been emptied, restarting peristalsis to clear out the guts. The stomach vibrations are thought to generate feelings of hunger. A rumbling stomach might feel odd, but it's perfectly normal.

Burps

Burps are probably slightly more welcome in polite society than farts, but only just. Also known as belches or eructations they are perfectly natural releases of gas from your stomach and oesphagus and are caused by gulping air down as you talk or swallow, or from the gas in your food and drink, especially carbonated drinks. Babies swallow lots of air as they feed which can cause them discomfort until it's released by burping. Not all of our gas gets burped back out, with much of it going all the way through your digestive system (swallowed air usually makes up around 25% of our farts). Cows burp a lot due to their ruminative digestive method, and their burps contain a high proportion of methane produced by methanogenic archaea from their gut.

Chapter 05:
Assorted Farty Trivia

The world's greatest farters

Mr Methane is *amazing*. He's easily the world's greatest living flatulist, appearing on TV shows clad in a green and purple superhero outfit and green mask, and possessed of an otherworldly control of his bowels. This man can fart for a whole, long, hot minute, blow candles out with his gas and shoot darts from his bum to pop a balloon (take a look at his eye-opening YouTube videos). But if you thought he'd be a loud, laddish clown you'd be surprised. He's tall (2m/6ft 7in) and middle-aged, British (from Macclesfield), his real name is Paul Oldfield and he's thoughtful, polite (he says 'motion' rather than 'poo'), with a rock-solid Macclesfield accent and a dry sense of humour.

Mr Methane discovered his talent (he told me that the talent found him) at the age of 15 when he was trying out yoga positions with his sister, and the contortions created huge farts. He carried on experimenting until his Dad walked in on him saying 'You'll have an accident if you carry on doing that.' His superskill remained underused for many years other than to win a few bets at school, and he focused on becoming an apprentice train driver until one night a friend took him to Macclesfield's Screaming Beavers club where he was dragged out of the audience to show off his unusual skills. After hooking up with the thoroughly indecent Macc Lads rock band (don't whatever you do listen to them at the office) who christened him Mr Methane, a glittering career unfolded, packed with international travel, prestige and notoriety.

Paul's technique is to take a poo around three hours before a performance to get the bowel pressure just right, then lie back to expand his sphincter and raise his diaphragm away from his bowels to 'breathe' the air in (he admits he doesn't *really* know how it works). It's all about relaxation and it's easy when lying on his back, but if he has to recharge standing up the air gets trapped in his transverse and ascending colon and feels sore. If you get a chance to book him or see him, you won't be disappointed. His rectal singing – as he farts along to the 'Blue Danube' – takes your breath away.

Le Pétomane

The world's greatest *dead* flatulist is undoubtedly the French star Joseph Pujol, born in 1857 in Marseilles and known as Le Pétomane (which roughly translates as The Fart Maniac). The son of a baker, he discovered an ability to ingest through his anus while bending over in the sea and simultaneously taking a deep breath – and felt a freezing-cold rush of water filling him up. After going into the army at the age of 20 he began to train himself to fart for as long as possible and vary the pitch so he could – near as dammit – sing with his bum.

By the mid-1880s he was working his fart act around France, and then in 1892 he auditioned for the world-famous Moulin Rouge nightclub in Paris. He was booked on the spot and within two years became the highest-paid performer in France. There are stories of him drawing box-office receipts of 20,000 francs per show. His act was packed with fart impressions starting with a newlywed bride on her wedding night (a shy little squeak) followed by the same bride a few months into marriage (massive flapping honk). He did animal impressions, played the flute with his bum, blew smoke rings out of his anus and ended his show with La Marseillaise before puffing out a candle with his farts (Mr Methane learned from the best). His shows were reportedly so funny that nurses were on hand (well, it was *claimed* that they were on hand) to deal with fainting and incapacitated audiences, and it all earned Joseph such a reputation that he socialised with Matisse and Renoir. His act did well until the Great War of 1914–18, after which he started a bakery and eventually died aged 88.

Roland

The greatest medieval farter was probably Roland, court minstrel to the 12th century King Henry II. His act ended with a dance including simultaneous performance of 'one jump, one whistle and one fart'. He was rewarded with Hemingstone, a manor house in Suffolk with over 40ha/100 acres of land, although it's believed that Henry III was less keen on guffing and took the whole lot back.

Great historical farts

History's most notorious fart is considered to be the one described by the historian Flavius Josephus in his *The Wars of the Jews,* written in AD 75. A boorish soldier farted at Jewish worshippers in Jerusalem during Passover, and this resulted in a riot and a stampede that killed 10,000 people. There is some debate as to whether the soldier merely exposed himself as an insult, but we're going with the fart.

In a more heroic vein, Herodotus tells of the unpopular Egyptian King Apries in 569 BC who sent his adviser Patarbemis to confront a rebel general, Amasis. Amasis replied to Patarbemis's confrontation with a fart to send back to Apries, and when the adviser returned home the King had his ears and nose cut off for failing in his task. The Egyptians were already feeling mutinous and were outraged at Apries' treatment of the adviser and joined the rebel general's revolt. It all resulted in the King's defeat and his eventual death years later as he tried to reclaim his throne.

Pythagoras was a Greek philosopher living around 570–495 BC, and founder of the mathematics-and-mysticism Pythagoreanism movement. He prohibited his followers from eating beans, and this is thought to be on anti-fart principles or the (completely understandable) worry that you might fart out your soul… but it's also possible that beans were taboo because they were associated with reincarnation.

Hitler is believed to have suffered from persistent and painful gastritis, and with it heavy flatulence, for which it has been suggested that he was prescribed all manner of medicines that may have made him psychotic. There's little dispute that he had serious and fascinating digestive problems, but I worry that the trouble with claims about his health and treatment – syphilis, monarchism (having only one testicle), methamphetamine habit (read Norman Ohler's captivating *Blitzed*) – are that they reduce the horrors of Nazi Germany to one man's supposed psychopathy, and trivialises or excuses the history of the Third Reich.

Benjamin Franklin was one of the founding fathers of the United States, and a renowned polymath. He was a newspaper editor, printer, inventor of both bifocals and the lightning rod, and postmaster general. He was also the author of a satirical essay written in 1781 while he was Ambassador to France entitled 'Fart Proudly' (also titled 'A Letter to a Royal Academy about Farting') stating that 'It is universally well known, that in digesting our common food, there is created or produced in the bowels of human creatures, a great quantity of wind.' He suggests the development of drugs that transform horrible farts into smells that are 'not only inoffensive, but agreeable as Perfumes.'

The whole piece takes a dig at learned societies in Europe that he believed were becoming pretentious and divorced from reality. As it's a satirical piece we can assume that rather than encouraging more research into farting, he was suggesting the opposite, using it as a way of saying that scientific endeavour in the era was a load of nonsense.

Now, listen. We all love a bit of satire but I think Franklin was just being a bit chippy. Let's take a look at the 'pretentious' scientific advances in just 1781, the year that he wrote his essay:

- Uranus discovered by Herschel and reported to the Royal Society
- Axons of a brain cell described by Fontanta
- A patent for the manufacture of coal tar granted
- Molybdenum isolated by Hjelm
- Messier catalogue published detailing 110 astronomical objects
- Méchain discovered 13 galaxies, one planetary nebula, one open cluster, one globular cluster and the dwarf galaxy NGC 5195, companion galaxy to the spectacular M51 Whirlpool Galaxy.

That's in just one year. 1781. IN EUROPE! Wind yer neck in, Benj.

Web of historical farts

The internet is full of wonderful video clips featuring famous farts dropped by mistake on live TV, and I must confess that every now and then (usually when I've got a crushing deadline looming), I'll spend an hour or so watching them. My personal favourites include the glamorous fitness coach leading three glamorous women in a series of stretchy floor exercises entitled, 'Love Your Body'. As she spreads her legs in preparation for an exercise she lets fly an excellent low-frequency guff before she and her team roll about the floor laughing. Also Whoopi Goldberg must be credited for taking responsibility for a real rasper as she listens to Bette Midler interviewing Clare Danes about how vitally important to the world her TV series is. Who knew that hubris had flavour?

Don't, whatever you do, search 'Babies scared of farts compilation'.

Farts in literature

The first recorded joke, written in 1900 BC, is about farts*, and many of our greatest writers use them in their work. And why not? They're perfectly natural yet offensive, rich material for insults, rude but not filthy, intimate but not sexual, and straddled between the unmatched concepts of biology and self-loathing. The Greek playwright Aristophanes' work *The Clouds* (423 BC) and *Frogs* (405 BC) both indulge the fart, and not just with a passing reference. He really goes to town, describing the thunderous sounds of 'pa-pa-pa-pax' at length. Seneca (4 BC–AD 65) toyed with farts, as did Horace.

Chaucer's *The Canterbury Tales*, written between 1386 and 1399, is one of those books that everyone hates when they're forced to read it

---

* It goes like this: 'Something which has never occurred since time immemorial: a young woman did not fart in her husband's lap.' I mean I get it, but is it *funny*?

at school… until they get to the filthy *Miller's Tale* with its squeamish parish clerk tricked into kissing a lady's hairy anus thinking it was her lips (he 'thought it was amiss for well he wist [thought] a woman hath no beerd'), red-hot pokers shoved up anuses and thunderous farting. Then everyone begins to enjoy it. *The Summoner's Tale* is preoccupied with how to share a fart among a group of friars.

Shakespeare deployed farts euphemistically and with surprising gravity. His classic fart reference is from the slapstick farce *The Comedy of Errors*: 'A man may break a word with you, sir; and words are but wind; Ay, and break it in your face, so he break it not behind.' Not his finest moment. But then nor was the rest of *The Comedy of Errors*.

My favourite literary fartfest belongs to the wonderful Jonathan Swift who simply couldn't stop himself from satirising people he thought were pretentious and self-important. In 'The Benefit of Farting Explain'd' (1722), he delights in 'such damned low stuff as farting', slagging off French farts and French turds for being pathetically weak in comparison to the English. The pamphlet was a parody of the Lord Bishop of Down and Connor's 'The Benefit of Fasting' and despite the fact that it verges on playground humour, his sheer enthusiasm makes it very funny. He writes under the pseudonym *Don Fartinando Puff-indorst, Professor of Bumbast in the University of Crackow* and the entire thing feels like a breath of fresh air. Take the title page alone:

> *'Translated into English at the Request, and for the Use, of the Lady Damp-fart of Her-fart-shire. By Obadiah Fizzle, Groom of the Stool to the Princess of Arsimini in Sardinia. Long-Fart (Longford in Ireland): Printed by Simon Bumbubbad, at the Sign of the Wind-Mill opposite Twattling-Street.'*

Swift suggests that there are five different species of farts, which are perfectly distinct from each other, both in weight and smell. First, *the sonorous and full-toned, or rousing fart*; second, *the double fart*; third, *the soft fizzing fart*; fourth, *the wet fart*; and fifth, *the sullen wind-bound fart*.

The Arabian Nights collection of Arabic folk tales has a fun, flatulent tale called 'How Abu Hasan Brake Wind', and many other wonderful writers have indulged the fart, including Dante, Mark Twain ('In ye heat of ye talk it befel yt one did breake wind, yielding an exceding mightie and distresfull stink, whereat all did laugh full sore...', as well as Rabelais, Ben Jonson, Victor Hugo and Balzac.

I had thought that the answer to the question, 'Can writing about farts be too dirty?' would be 'Never'. But I was wrong, James Joyce got down and dirty – yay, filthy-dirty-filthy with farts in some of the most vulgar love letters you could ever imagine reading. I was actually shocked, which is saying something, coming from a guy who's spent most of the last year typing the work 'anus' into search engines. (Don't. Just don't. I did it so you don't have to.) *That's* when they can be too dirty.

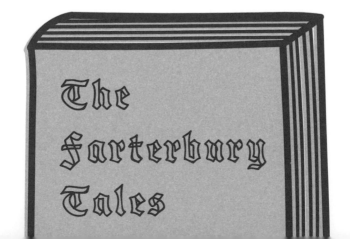

A small dictionary of fart slang

The fertile linguistic field of flatus has been enthusiastically ploughed by writers great and small, with varying levels of success. The best ones leave you bewildered for half a second.

Top ten euphemisms

1. Get out and walk, Donald
2. Brown thunder
3. The toothless one speaks
4. Butt dumpling
5. Mud duck
6. Greaser
7. Thunder from down under
8. Stepped on a frog
9. Shart
10. Release the hounds

The beautifully baffling ten

1. Air tulip
2. Answering the call of the wild burrito
3. Barking spider
4. Fluffy
5. String of pearls
6. Roast the jockeys
7. Grundle rumble
8. Let Polly out of jail
9. Shoot a bunny
10. Cockney cheers

Ten kid-friendly crowd-pleasers

1. Cheeky squeaky
2. Great big flowery woof woof
3. Step on a duck
4. Bumsen burner
5. Benchwarmer
6. Beefy eggo
7. Great big blast of joy
8. Bottom burp
9. Cut the cheese
10. Great brown cloud of fun

Ten less fragrant euphemisms

1. Arse flapper
2. Crack a rat
3. Butt bongos
4. Gravy pants
5. Blasting the arse trumpet
6. Fecal fumigation
7. Exhume the dinner corpse
8. Arse trumpet
9. Turd tremor
10. Heinous anus

The international fart dictionary

1.	**Pet** *(hence Monsieur Pétomane)*	French
2.	**Furz** *(vulgar: Scheißer)*	German
3.	**Scoreggia**	Italian
4.	**Perdet**	Russian
5.	**Brodler**	Walloon
6.	**Rhech**	Welsh

(e.g. 'Fel rhech mewn pot jam': 'Like a fart in a jam jar' or 'useless')

7.	**Jamba**	Swahili
8.	**El pedo**	Spanish
9.	**Apaan vaayu**	Hindi
10.	**Durta**	Arabic

The Shorter Oxford English Dictionary Definition

Fart /faːt/ *v. & n. course slang* • *v.intr.* **1** emit wind from the anus. **2** (foll. by *about, around*) behave foolishly; waste time. • *n.* **1** an emission of wind from the anus. **2** an unpleasant person. [Old English (recorded in the verbal noun *feorting*) from Germanic]

INDEX

A

air, swallowing 109, 114, 127
aircraft 110
alpha-galactosidase 111
amino acids 19, 51, 112
ammonia 34–5
anaerobic digestion 11,
96–103
animals 30–2
antibiotics 111
anus 46, 47, 77, 83–5, 87,
116
asparagus 61–2

B

bacteria: in colon 11–12,
46, 68–70
   cross-contamination 123–4
   faecal transplants 74
   in farts 17–18
   gas production 18
   metabolism 11, 80–1
   sewage plants 93
bananas 51, 117
beans 49, 62–3, 109, 112
beer 112
beetroot 60–2
Bernoulli's principle 84,
86–7
biogas 94–5
body builders 52
brassicas 50, 112

breath, farty 79
Bristol Stool Form Scale
72–3
burps 79, 127
buses 94–5

C

cabbage 50, 114
carbon dioxide 12, 37–39,
51, 65, 110
charcoal 111, 112–13
chewing 40, 109, 114
chyme 43, 46, 76–7
Coandă effect 86–7
cocktail, Dutch fart 63
colon 14, 39, 45–6
   bacteria in 11–12, 46,
   68–70
   diverticulosis 79
   peristalsis 42, 77
   releasing stress 116–17
cows 31, 65, 127

D

dairy products 52, 109, 117
digestion 14, 39–47, 71–77
dimethyl sulfide 20, 22
disease 118, 122–3
diverticulosis 79
duodenum 43, 105
Dutch fart cocktail 63
Dutch ovens 18

E

embarrassment 118–21
energy 37–9, 80–1, 94–5
enzymes 41, 42, 44, 76–7

F

faecal transplants 74
famous farters 116, 129–31
fart machines 96–103
fatty foods 51, 112
fibre 39, 46, 108–9
fizzy drinks 110, 114, 127
flammability 25
flatulists 116, 129–31
food: digestion 39–47, 75–7
   fartiest foods 49–63
   increasing wind 114–17
fructans 49, 52, 117
fruit 50–1
fungi 70

G

garlic 49, 112
gases 12–13, 18–23
gastroscopy 105–7
giardiasis 117
grains 50
gum, sugar-free 53, 110

H

handwashing 121–4
hiccups 125
historical farts 132–5
hot farts 80–1
hydrogen 12–13, 37, 65
hydrogen sulfide 20, 22, 24,
51, 52, 67

I

increasing wind 114–17
indole 20, 23
inulin 48–9, 55, 57
irritable bowel syndrome
(IBS) 74, 79, 111

J

jars, collecting farts in 26–7
Jerusalem artichokes 48–9, 54–7, 58–9, 61–2

L

lactose 52, 109
large intestine see colon
leaf blowers 86–7, 89
light 37–39
literary farts 135–7

M

meat 51, 112
men's farts 24–5
methane 13, 25, 65, 93, 94–5, 96, 127
methanethiol 20, 23, 24
methyl thiobutyrate 20, 23
miasma 119, 121–2
Mr Methane 129–30
mushrooms 52

N

noises 83–5

O

oesophagus 41–2, 75, 109
olfaction 20–1
oligosaccharides 11, 49, 109
onions 49, 112
oranges 51
oxygen 11, 37–8, 65

P

pain 79
pants, fart-filtering 113
pasta 52
peppermint tea 111
Pepto-bismol 112
peristalsis 42, 75–7, 127
Le Pétomane 116, 130–1
photosynthesis 37–39
PillCam 105–6
plants 37–39
pond weed 39
poo 15–18, 46–7, 71–4, 77
potatoes 52, 117
probiotics 111
protein 51, 52, 112

R

raffinose 49, 109
recipes 58–63
rectum 46, 47, 77, 83–4
reflexes 125–6
rocket fuel 58–9
rumbling stomach 127

S

saliva 43
sausages 64–5
scatole 22, 25
septic tanks 93
sewage plants 92–5
simethicone 113
slang 140–2
small intestine 46, 78–9
smells 15, 20–7, 30–2, 114
sneezing 128–9
sorbitol 55, 112
sounds 85–7
sphincter, anal 47, 77, 83–5, 87–9
stink bombs 34–5
stomach 42–3, 76, 128
sunlight 37
swallowing air 109, 114, 127
swallowing food 41–2

T

thiols 20
transplants, faecal 74
'trapped wind' 65
trimethyl amine 20, 23

V

vegetables 48–50, 109, 112
videos 81, 134–5

W

washing hands 121–4
wheat 117
women's farts 24–5

Y

yawning 126

Acknowledgements

Huge thanks to Andrea Sella (Clever Fella), my gorgeous friend who's always listed on my call sheets as 'Overworked chemistry fruitcake'. He inspired me to explore food using science, and to find new ways of telling the stories behind the stuff we put into our mouths. Several years ago we came up with the idea of a Fartology science stage show, and as we watched audiences wetting themselves with laughter whilst learning about complex multidisciplinary science, we realised we had created something special. The UK should be immensely proud of its thriving science communication scene, and I'm enormously grateful to have been given a platform to create merry chaos and spread knowledge by a huge variety of enlightened organisations from Cheltenham Science Festival to Butlins (with a special thank-you to Mike Godolphin).

A lot of fart enthusiasts have helped in many different ways with this book: Heather Fitzke, Mark Lythgoe, Mohammed Saddiq, Chris Clarke, Hugh Woodward, Charlie Torrible, Theo Blossom, Philip Woodland, Alex Menys, Steve Pearce, Paul McKnight, Ewan Bailey, Brodie Thomson, Eliza Hazlewood, Jan Croxson, Borra Garson, Louise Leftwich, Nicholas Caruso, Daniella Rabaiotti, Shelli Martelli and Gina Collins.

Thanks so much to Sarah Lavelle for letting me sully the good name of Quadrille with my weird obsession with farts (and for letting me fart in the faces of half the team), to Harriet Webster and Kathy Steer and everyone at Quadrille who worked on the book, and to Luke Bird for his fantastic design.

Thanks to my delicious girls: Daisy, Poppy and Georgia. Sorry for all the dreadful, dreadful smells you've had to endure in the making of this book.

And lastly, thanks so much to the brilliant audiences who've come along to my shows and laughed their pants off along with the Gastronaut team whilst we've explored some utterly revolting science live on stage. I LOVE you people!